UNIFICATION OF FUNDAMENTAL FORCES

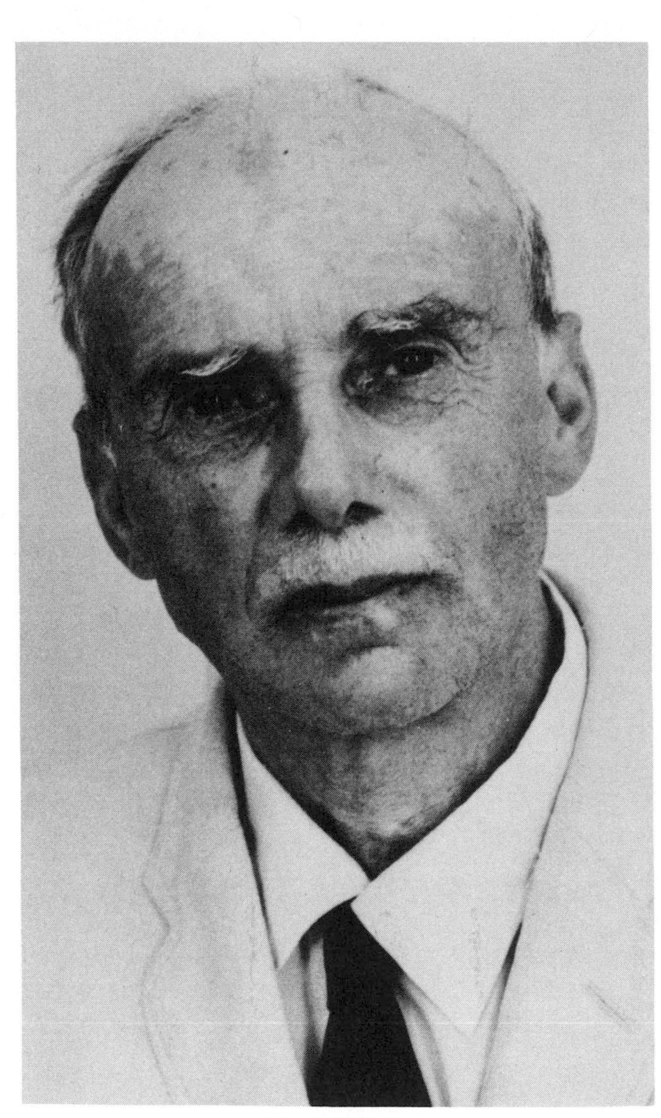

Paul Dirac

UNIFICATION OF FUNDAMENTAL FORCES

THE FIRST OF THE 1988 DIRAC MEMORIAL LECTURES

ABDUS SALAM

Imperial College, London
and International Centre for Theoretical Physics, Trieste, Italy

Lecture notes compiled by
Jonathan Evans and Gerard Watts

CAMBRIDGE UNIVERSITY PRESS

Cambridge

New York Port Chester Melbourne Sydney

Published by the Press Syndicate of the University of Cambridge
The Pitt Building, Trumpington Street, Cambridge CB2 1RP
40 West 20th Street, New York, NY 10011, USA
10 Stamford Road, Oakleigh, Melbourne 3166, Australia

First published 1990

Printed in the United States of America

Library of Congress Cataloging-in-Publication Data

Salam, Abdus, 1926–
Unification of fundamental forces: the first of the 1988 Dirac memorial lectures/Abdus Salam; lecture notes compiled by Jonathan Evans and Gerard Watts.
p. cm.
ISBN 0-521-37140-6
1. Grand unified theories (Nuclear physics) 2. Quantum theory. I. Dirac, P. A. M. (Paul Adrien Maurice). 1902– . II. Evans, Jonathan (Jonathan M.) III. Watts, Gerard. IV. Title. V. Title: Dirac memorial lectures.
QC794.6.G7S25 1989
530.1′2–dc20 89-15861
CIP

British Library Cataloguing in Publication applied for.

ISBN 0-521-37140-6 hard covers

CONTENTS

FOREWORD

John C. Taylor

University of Cambridge

From time to time, science succeeds in unifying apparently diverse sets of phenomena. These unifications provide some of the most impressive achievements in the sciences. Unification, in this sense, means understanding how apparently different effects are really aspects of a single underlying thing. In the nineteenth century, for example, electricity and magnetism were unified. They are different, but they are intimately interconnected, and in general situations it is impossible to imagine one without the other.

In physics, one of the unifications of the present century has been that of electromagnetism with the weak force. These are apparently totally different. Electromagnetism ranges across any distance, from atomic to astronomical. The weak force, on the other hand, operates deep within the atomic nucleus (for example) to produce radioactive decay. This unification was mainly the work of three men, Sheldon Glashow, Abdus Salam and Steven Weinberg. They made use of earlier theoretical ideas of two British physicists Peter Higgs and Tom Kibble. Finally, a Dutch theoretician, Gerard t' Hooft, put the equations into a form in which any competent physicist could handle them in standard ways. The experimental implications of all this are at present being tested with the new high-energy colliders at CERN, Fermilab and Stanford. A whole new branch of physics is unfolding, just as the science of electromagnetism did in the last century.

In this volume, Abdus Salam, one of the discoverers of the electro-weak unification, writes about unification in physics in the past and present, and describes hopes for the future. The whole enterprise takes place within the framework of quantum mechanics, of which one of the main architects was the great English physicist Paul Dirac. The volume is based upon a Dirac Memorial Lecture given at Cambridge University in 1988.

Salam and other authors of the electro-weak theory operated squarely within the framework of quantum theory, which was laid down in the first half of this century. Two of the great founders of quantum theory were Werner Heisenberg and Paul Dirac. A lecture by each of these is included in this volume. These, like Salam's lecture, provide fascinating insights into the thoughts of creative theoretical physicists.

UNIFICATION OF FUNDAMENTAL FORCES

THE FIRST OF THE 1988 DIRAC MEMORIAL LECTURES

Abdus Salam

1. INTRODUCTION

I am honoured to speak about P. A. M. Dirac whom we all loved and whom I so greatly admired. I am also glad to see so many friends in the audience. As an old Johnian myself, I would particularly like to mention Sir Harry Hinsley, the Master of St. John's (Dirac's College). Sir Harry is an eminent historian. To him I shall address my remarks, so as to assure you all that you will be spared as many technical details as possible.*

Paul Adrian Maurice Dirac was undoubtedly one of the greatest physicists of this or of any century. In three decisive years,† 1925, 1926, and 1927, with

* Professor Stephen Hawking, a worthy successor to Newton and Dirac, in his admirable new book (*A Brief History of Time*: Bantam books, 1988) says that he was told that if a popular science book contains one formula, the sale of the book becomes halved. I shall try to use no formulae at all (except in the footnotes) – apart from Einstein's notorious equation, "$E = mc^2$", and powers of 10 – which are inevitable in this subject.

† I owe this remark to Victor Weisskopf.

three papers, he laid the foundations, first of the Theory of Quantum Mechanics, second of the Quantum Theory of Fields, and third – with his famous equation of the electron – of the Theory of Elementary Particles. (In the course of this lecture, I shall explain the relevant concepts of the Quantum Theory of Fields and the Dirac equation for the electron.) When one met Dirac, one could see the complete and utter dedication of a great scientist. One could feel with him the pleasure of scientific creation at its noblest, and the highest personal integrity. He had a childlike simplicity about him. His lucidity and clarity of thought made him a legend. He was undoubtedly one of the greatest human beings I have had the privilege of meeting during my life.

For those of you who never met Dirac, I would like to quote from a reporter's article which was written about him in 1934 by a newspaperman at the University of Wisconsin. It says:

> I have been hearing about a fellow they have up at the University this Spring. A mathematical physicist or something they call him, who has been pushing Sir Isaac Newton, Einstein and all the others off the front pages. His name is Dirac and he is an

Englishman. So the other afternoon I knocks on the door of Doctor Dirac's office in Stirling Hall and a pleasant voice says 'Come in'.

And I want to say here and now that this sentence 'Come in' was one of the longest ones emitted by the doctor during our interview.

I found the doctor a tall, youngish man and the minute I sees the twinkle in his eyes I knew I was going to like him. He did not seem at all to be busy. When I want to interview an American scientist of his class, he would blow in carrying a big briefcase, and while he talked he would be pulling lecture notes, proofs, books, reprints, manuscripts and what-have-you out of his bag.

Dirac is different. He seems to have all the time that is in the world and his heaviest work is looking out of the window.

'Professor,' says I, 'I notice you have quite a few letters in front of your last name. Do they stand for anything in particular?' 'No' says he.

'Fine' says I, 'now will you give me the lowdown on your investigations?' 'No' says he.

I went on. 'Do you go to the movies?'. 'Yes' he says. 'When?' 'In 1920.'

Dirac described his own life in physics in a lecture he gave in Trieste in 1968.* I was reading it last night and I came across some parts which seem particularly relevant to the theme of unification of fundamental forces. Dirac describes in this lecture how he got his ideas, particularly the distinction between two methods of investigation in theoretical physics.

According to Dirac, first one may try to make progress by searching for a mathematical procedure for the removal of inconsistencies† which may be

* Reprinted in its entirety on p. 125 of this volume.

† The inconsistencies Dirac had in mind referred to the so-called infinity problem. The problem was that all higher-order calculations in quantum field theories yield the result logarithm infinity ($\log \infty$). Dirac, before World War II, had suggested that all these inconsistencies can be lumped together into an unobservable "renormalisation" of the electron rest-mass. F. J. Dyson, in 1949, showed that Dirac's conjecture was right for electrodynamics and that all inconsistencies could be incorporated into two "renormalisations", one for the electron rest-mass and the second for the electron charge.

Dirac, although he had suggested the idea in the first place, disliked these so-called renormalisable theories. He kept hoping that this last vestige of inconsistency would also disappear eventually, with the final theory emerging as pure as driven snow. In the version of unified superstring theories, where gravity is also united with gauge forces (see p. 75), we believe we are in sight of Dirac's goal.

present in a physical theory like electrodynamics of electrons – I shall mention some of Dirac's own work in this connection later and the relevance of his ideas today. Second, one may try to unite theories that were previously disjoint. Dirac says that this second method had not proved very fruitful. He was perhaps speaking from the exasperation felt by many of his generation with attempts that had been made, particularly by Einstein, to unify fundamental theories and which had met with scant success.

In contrast, our generation has been mainly concerned with this second method, and my talk will almost entirely be devoted to it.*

* F. J. Dyson in his beautiful recent book *Infinite in all Directions* (Harper and Row, Cornelia and Michael Bessie books, 1988) has this to say about the second of Dirac's ideas: "Now it is generally true that the very greatest scientists in each discipline are unifiers. This is especially true in physics. Newton and Einstein were supreme as unifiers. The great triumphs of physics have been triumphs of unification. We almost take it for granted that the road of progress in physics will be a wider and wider unification bringing more and more phenomena within the scope of a few fundamental principles. Einstein was so confident of the correctness of this road of unification that at the end of his life he took almost no interest in the experimental discoveries which were then beginning to make the world of physics more complicated. It is difficult to find among physicists any serious voices in opposition to unification".

2. The unifying concepts of physics in the past

I shall start by giving you an idea of the variety of unifying concepts that have been used in physics from the beginning. Figure 1 outlines the history of the unification of physical theories, and by the end of this lecture I shall have reached the 'Theory of Everything' (TOE) at the bottom of the figure, which is where the fun lies at the moment. Most of my talk, however, will be devoted to describing today's 'Standard Model' of particle physics, which has emerged as a consequence of our generation's efforts to determine what the elementary entities are and to unify some of the forces of nature between these elementary entities.

The Galilean principle

The first name I would like to mention in this context is that of Al-Biruni who flourished in Afghanistan a thousand years ago. One might not think of modern Afghanistan as a likely place where high class physics could be done. Al-Biruni however, to my knowledge, was the first physicist to say

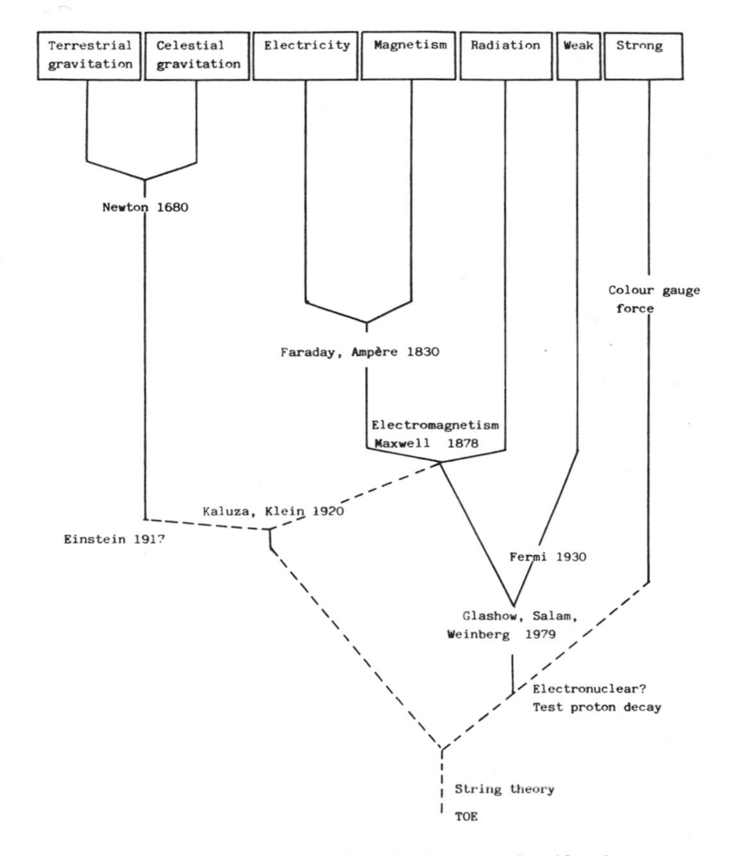

Fig. 1 A diagram showing the history of unification of physical theories.

explicitly that physical phenomena on the Sun, Earth and the Moon obey the same laws.*

This deceptively simple idea is the basis of all of science as we know it. This was independently stated and demonstrated by Galileo six hundred years later. Galileo used his telescope (imported from Holland) to observe the shadows cast by mountains on the moon. By correlating the direction of the shadows with the direction of the sunlight, he was able to assert that the *laws of shadow making were the same on the Moon as on the Earth.* This was the first demonstration of the fundamental principle – now known as the 'Galilean Symmetry'† – which asserted the universality of physical laws.

* This was one of the "arguments" which occupied the minds of men in the Middle Ages. Clearly, there would be no universal Science if the basic laws depend on where we happen to be situated in the universe, or when the experiments were done.

† Suprisingly, one consequence of this symmetry is the conservation of momentum; that is to say, total momentum in any initial state equals the total momentum of a system in the final state, irrespective of what interaction takes place in between the initial and final stages. Symmetry principles almost always give rise to conservation laws, of which the above is an example.

Isaac Newton and the unification of terrestrial and celestial gravities

The next person to mention in this context is Isaac Newton. Around 1680 he asserted that the force of "terrestrial" gravity (which makes apples fall to the ground, and which in Newton's view was a universal force) was the same as "celestial" gravity (the force which keeps planets in motion around the Sun). Such a force is long-range. Its effects can be felt at any distance, though attenuated by the square of the distance between the two "gravitating" objects concerned.

Newton introduced a new fundamental constant of nature, G, which characterises the strength of the gravitational force. The constant G is very small in magnitude*:

Gravity is always attractive in contrast to other forces of nature which, as we shall see, can be

* $G = 6.670 \times 10^{-11}\ \mathrm{Nm^2\ kg^{-2}}$. As we shall see later, when Planckian units are used, this number works out to be $10^{-40}(\text{proton mass})^{-2}$. For those who find it difficult to think in terms of powers of 10, it is perhaps useful to have the following glossary:

10^3 = one thousand	10^{-3} = one thousandth
10^6 = one million	10^{-6} = one millionth
10^9 = one billion	10^{-9} = one billionth
10^{12} = one trillion	10^{-12} = one trillionth

repulsive as well as attractive. This gives gravity the edge in that the force always adds up.

Faraday and Ampere; *The unification of electricity and magnetism* (1820*s to*1830*s*)

The next unification of fundamental forces was postulated some 150 years later. I wish to recall Faraday and Ampere in the context of *electromagnetism* – the 'force of life' (so-called because all chemical binding is electromagnetic in origin, and so are all phenomena of nerve impulses).

Before 1820, electricity and magnetism were regarded as two distinct forces. Faraday and Ampere, in the greatest unification of modern times, were the first to show that electricity and magnetism were but two aspects of one single force – *electro-magnetism.* If one considers an electrically charged object, an electron for example, and places it on a table so that it is stationary, one could detect (by placing another electron near it) an electric force of repulsion. But as soon as the first electron is moved, one will also encounter a magnetic force which was not there before. This force can be detected by its effect on a compass needle placed near the moving electron.

It is thus an *environmental factor* which distinguishes between electricity and magnetism – *namely whether or not the electric charge is in motion or not. This is the essence of the unification of electricity and magnetism.* (To anticipate, we shall come upon a similar environmental factor, the temperature of the early universe, in the context of another unification – that of electromagnetism with the weak nuclear force.)

This unification of the two disjoint fundamental forces of electricity and magnetism was the basis of electrical current technology of the nineteenth century which depended on the generation of electric currents by rotating a coil of wire between the two poles of a magnet. This is the basis of the electric motors and electric dynamos which led to the electric power stations. These followed on the remarkable unity of two of the disparate forces of nature. I do not believe that simple tinkering would ever have produced the quantitative basis of this development.

Maxwell and the unification of electromagnetism with optics

Classical electromagnetism found its culmination fifty years later in the work of Maxwell, who showed that if an electric charge is *accelerated* (i.e. its speed changes or there is a change of direction), it would emit energy in the form of electromagnetic radiation (radio waves, heat waves, light rays, X-rays, gamma rays which differ from each other in respect of their wave lengths only). This miraculous unification has been the basis of technology of the twentieth century, with radio, television and X-rays dominating our lives.

Maxwell had unified optics with electromagnetism. What is theoretically most remarkable about this unification is that he only had one number to guide him. Using rather crude apparatus,* he verified that the speed of light could be expressed in terms of two known constants (which described *electric and magnetic properties of the vacuum*), as predicted by his theory. This was an "indirect" confirmation of his ideas. He was unfortunate in that he died at the early age of 48 and did not live to

* One of my fondest recollections as an undergraduate is that of working with Maxwell's equipment in the Old Cavendish to verify the relation mentioned in the text.

Fig. 2 Heinrich Hertz (at top), in 1888, produced electromagnetic radiation by accelerating electric charges, thus confirming Maxwell's theory unifying electromagnetism and optics.

see electromagnetic radiation produced by accelerating electric charges – demonstrated by Hertz in Germany some ten years after Maxwell's death.

Einstein's unification of space and time and generalisation of gravity

This finally leads us to Einstein who was responsible for a number of far-reaching unifying ideas. His Special Theory of Relativity (1905) places space and time on an equal footing. One consequence of this work was the time dilatation formula which says that the faster a body moves, the longer its life is – as observed by a stationary observer. This particular phenomenon can be observed any day with the greatest precision possible at the CERN Laboratory at Geneva, where a particle like the muon with a definite lifetime (as measured by an observer riding with the particle) appears to us, the stationary observers, to live longer and longer as its speed approaches the velocity of light. Thus, the secret of longevity is to stay in motion!

Another consequence of Einstein's special relativity theory is the well-known *relationship between mass and energy* embodied in the famous equation

$E = mc^2$ where c is the velocity of light, m is the mass of the moving particle and E is its energy.*

Einstein and Gravity Theory

Einstein went further in his general theory of relativity (1915). He geometrised physics in the sense that in his theory, the curvature of space and time determined gravity. Curvature is a geometric notion while gravity is one of the fundamental forces of nature. Einstein, by a stroke of genius, equated these two and thereby accomplished the *geometrisation* of physics.

This idea was new to most people in Einstein's day, but he was not completely without precedent in his thinking.† Gauss, around a hundred years earlier, had also had the idea that space might be curved. He actually made experiments to test this. What Gauss did was to put observing stations on the tops of three mountains, each a few miles apart.

* In this lecture I shall use the words mass and energy interchangeably, i.e. in the formula $E = mc^2$ we choose units of time and distance such that the velocity of light $c = 1$.

† Einstein's theory explained a small discrepancy which had arisen with Newton's theory (in the determination of the orbit of Mercury). This was one of the reasons for its eventual general acceptance.

He measured the angles made by light rays that were reflected from one station to the other to make up a triangle. If the angles of the triangle had added up to less (or to more) that 180°, he would have shown that space was curved. He found a null result – the angles he measured did add up to 180°. We now know this was because he was working with distances of only a few miles rather than with stellar distances where the curvature of spacetime is manifest.

Friedman and Hubble: Penzias and Wilson's radiation and the Big Bang

Einstein's theory of gravity revived the idea that space and time were curved. The next steps were taken by the Russian astrophysicist Friedmann, who considered the overall structure of the universe and found that an expanding universe could arise as a solution of Einstein's equations. This was experimentally confirmed by Hubble, who discovered that distant galaxies are receding from us, precisely in accordance with these ideas.

In 1965, Penzias and Wilson accidently found a radiative background with temperature of 3° kelvin (−270 °C) which apparently filled all of space. This was interpreted as radiation which had arisen 2×10^5 years ($\approx 10^{13}$ seconds)* after the Universe began its life. The expansion of the Universe had cooled the radiation between the time of its production and its detection today (10^{10} years after "time" began).†

Extrapolating backwards from 2×10^5 years, we would (speculatively) arrive at a moment when the universe began, the moment of the so-called Hot Bang.

Unification of gravity and electromagnetism

After the successes of General Relativity Theory and its explanation of gravitation in terms of curvature of space and time, Einstein started to wonder whether there was a connection between gravity and electromagnetism – in particular, could electromagnetism also be looked upon as a *geometrical* property of spacetime, thereby uniting the two forces of nature. Both these forces obey the same inverse square law, although their strengths at comparable distances are vastly different.

A number of years earlier, Faraday had con-

* 1 year $\approx 3 \times 10^7$ seconds.

† A gas, when it expands, cools. Likewise, the *expanding* universe cools.

ducted experiments to find a unity between electricity and (Newtonian) gravity. Using what he thought were large enough weights in free-fall, he hoped to detect an electric current, as shown by a nearby galvanometer.

Faraday found no effect. Writing in his diary, he noted:

> Here end my trials, for the present. The results are negative. They do not shake my strong belief of the existence of a relation between gravity and electricity, though they give no proof that such a relation exists.

Faraday was an extremely careful experimenter and he recorded his results, independently of his theoretical prejudices. Nowadays people believe that the effect Faraday sought should exist, and even that it may have been seen, but that it will show up most convincingly in cosmological experiments where the weights to be "dropped" should be of *stellar* sizes.

Such a unification of general relativity and electromagnetism was a dream which Einstein lived with and worked on for much of his later life. He spent 35 years on this problem and in the end, we

Fig. 3 Faraday attempts to unify electricity and gravity. (Reproduced with permission from A. de Rújula.)

believe, did not succeed. This may explain why Dirac in 1968 remarked negatively about the possibility of unification of fundamental forces.

Kaluza–Klein and extra spacetime dimensions

One attempt in this direction, however, deserves to be mentioned. This is the attempt of Kaluza, elaborated later by Klein. Kaluza sent a paper to Einstein – in 1919 I believe – in which he took the bold step of proposing that spacetime with five dimensions should be considered for purposes of (geometrically) uniting electromagnetism with gravity. He verified that the curvature corresponding to the extra (fifth) dimension gives the electromagnetic force, just as the curvature in the three familiar space dimensions plus time gives rise to gravity.

What are five dimensions? Imagine looking at a pencil from a long distance off. From a distance it looks like a thin line – one dimensional – and one does not notice that it is really a small cylinder with a two-dimensional surface. In the same way, five dimensions can look four-dimensional if the extra fifth dimension is tiny. (Klein, in fact, postulated that the extra dimension should be curled down to a length of around 10^{-33} cm (the Planck length)* in order that the curvature corresponding to the fifth dimension should correspond to the "correct" magnitude† of electric charge (i.e. the unit charge on the proton.)

Kaluza sent his paper to Einstein to be submitted for publication. Einstein (though he liked the idea of an extra unseen dimension in the beginning) had his doubts. The result was that he was responsible for delaying its publication by two years. Kaluza felt so unhappy that he left off fundamental physics and apparently, started work on the theory of swimming. The moral to me is this: if you have a reasonable idea, don't send it to a great man; publish it yourself.

* The Planck length is a very important concept in the quest for unification of all four known fundamental forces. The Planck "length" corresponds to the length scale at which gravity must be treated quantum mechanically, or in other words, the scale at which the final unification of the strong, weak, electromagnetic and gravitational forces should occur. To experiment with such very short distances, a great deal of energy is required. The "Planck" energy, needed to investigate these minute distances, is of the order of 10^{20} proton masses.

† In other words, the Planck length is the distance at which the force of gravity between two protons equals the electric force between the two.

I will anticipate my later remarks, and mention that there has been a modern revival of Kaluza–Klein-like ideas and that this is crucial for our candidate Theory of Everything (TOE). In superstring theory, for example, we start by postulating *ten* dimensions. Of these ten dimensions, four are the familiar spacetime dimensions while the rest are compactified to small Planck sizes of 10^{-33} cm, in order to accommodate electromagnetism *as well as* the nuclear forces. This way all these forces could combine with gravity into *one* fundamental force.

3. THE CONCEPT OF ELEMENTARITY AND NUCLEAR FORCES

Up to this point I have talked mainly about gravity and electromagnetism, which were the only known physical forces up to the early part of this century.

I should now talk of the two further forces, the two nuclear forces – which were discovered during this century and which Einstein (and Dirac for that matter) consistently ignored.

The nuclear forces are of two types, the so-called "weak" and "strong". Before I discuss them, knowledge of the elementary entities which interact via these forces will be required.

The concept of elementarity of *matter* is something that has evolved as time has gone by. Of the four Greek "elementary" entities, three (earth, water and air) could be called "elementary" entities of *matter*, while the fourth (fire) represented a *force*. If one was working in 1888, say, one would think of atoms as the fundamental elementary particles of the day, and *chemistry* as the science of *elementary particles.*

The proton and the neutron doublet, the quarks, and the strong nuclear force

Rutherford's experiments (done around 1910)

showed that atoms are not elementary; rather they consist of a small, dense, central nuclei, (some 10^{-12} cm in radius) surrounded by orbiting electrons. What Rutherford did was to scatter high-energy helium atoms, stripped of their electrons so that they were positively charged, off ordinary matter. He discovered, to his amazement, that some of them came right back at him. It was so astonishing he said it was as if one had shot bullets at a piece of paper and some of them had bounced back. This could only be explained by postulating that there was a small hard, dense, central nucleus inside an atom which contained nearly the whole of the atomic mass. Calculations showed that this nucleus was some 1/10000 times smaller than the atom itself.

Later still, the 1932 researches of Chadwick and Joliot-Curie revealed that the nucleus itself is made up of yet smaller particles: the protons (p) and the neutrons (n), each 10^{-13} cm in radius. These can naturally be regarded as a pair – a *doublet* of objects, *with nearly the same masses.* The proton differs from the neutron in being electrically charged:

$$\begin{pmatrix} p^{+} \\ n^{0} \end{pmatrix}$$

Although the proton is positively charged (which

the neutron is not), there are certain *symmetries* between their "strong" interactions. For example, pp force = pn force = np force = nn force, when electromagnetism *is not taken into account*. (This is a good starting assumption since, at distances of the order of 10^{-33} cms, the pn force is some 1000 times as strong as the *electromagnetic* force between two protons.)

There is one crucial fact about these strong nuclear forces – their short-range character. If protons and neutrons are at a distance larger than 10^{-13} cm, the strong nuclear force is essentially zero. All that remains at distances larger is the electromagnetic force (between protons) and, of course, the universal force of gravity.

Quarks as the elementary units of which protons and neutrons are made

Later research (by Hofstadter around 1956) has made it plausible that protons and neutrons are themselves not elementary point particles but have a definite size and are therefore composites. Today we believe, as the result of findings at SLAC (Stanford Linear Accelerator Centre), that they are made up of still smaller objects called quarks, which may themselves be elementary and point-like – an idea introduced into the subject (in 1963) by Gell-

Mann and Zweig. There are six types of quarks distinguished from each other by a property called "flavour". This "flavour" carries whimsical names like up-flavour (u), down-flavour (d), charmed-flavour (c), strange-flavour (s), top-flavour (t) and bottom flavour (b).

The six quarks apparently divide into three doublets (u, d); (c, s) and (t, b), i.e. three *similar "generations"* of the quark *"family"*. Why are there three generations? Does this mean that quarks are not really elementary entities, but that the second and third generations are related in some simple way to the first generation?*

A proton is made up of two up-quarks and one down-quark: p = (u, d, u). A neutron is made up of two down-quarks and one up-quark: n = (d, u, d).

Leptons and the weak nuclear force

Corresponding to the six quarks, there are six "*lighter*" particles – the so-called leptons,† which

* New experiments which will be conducted at the electron–proton accelerator HERA, under construction at present at the DESY (Deutsches Elektronen-Synchrotron) Laboratory in Hamburg (and due to begin experiments in 1991), will attempt to determine the sizes of quarks. The projectiles in this case will be high-energy electrons and the method used will be essentially the same as that of Rutherford.

† The mass of the electron is 1/2000 times that of the proton.

also divide into three doublets (ν_e, e), (ν_μ, μ) and (ν_τ, τ). Here ν_e, ν_μ, and ν_τ are three neutrinos while the charged objects e^-, μ^-, and τ^- each carry an electric charge equal in magnitude but opposite in sign to that of the protons. The existence of ν_e was postulated by Pauli in 1930. It was discovered by Reines and Cowan in 1955.* The (left-handed members† of these) doublets exhibit the so-called "weak nuclear force" in their mutual interactions – a force with a range shorter than 10^{-16} cm. We come back to the "weak" nuclear force later.

Dirac equation for elementary entities; intrinsic spin and handedness

This brings us to Dirac's famous equation which was first written down in 1927 for electrons. This equation can equally describe elementary entities like free electrons or free quarks or even free composite entities like protons and neutrons (free, when they are not interacting with similar objects).

The first important point about the equation is that while the only inputs to it were Einstein's special theory of relativity and quantum mechanics,

* The mass of the neutrino is very small, and may be zero; its precise value has not yet been determined.

† Left-handed particles are defined in the next section.

the equation did succeed in describing the correct "intrinsic spin" as well as the "handedness" of electrons.*

In quantum mechanics, "*intrinsic spin*" is a very important concept. It signifies that every chunk of matter (or energy) behaves like a rotating top and therefore possesses intrinsic spin.

Next: *helicity or "handedness*" is defined relative to a particle's direction of motion as the component of the intrinsic spin along or opposite to this direction.

The only possible values helicity can take are 0, $\pm 1/2$, ± 1, $\pm 3/2$, ± 2 ... (in units of $\hbar$, the Planck constant). The corresponding intrinsic spins† are 0, $\hbar/2$, $\hbar$, $3\hbar/2$, $2\hbar$ ....

Where does the intrinsic spin come from? Dirac's answer was: from uniting Einstein's special relativity with quantum mechanics!

Dirac's equation describes only spin $\hbar/2$ objects (with just two helicities $1/2\,\hbar$ and $-1/2\hbar$). One may think of these as rotating tops – either spinning

* The equation contained one parameter, the mass of the electron; when the equation is modified for the free proton or the free neutron all one has to do is to change the mass parameter appropriately.

† A distinction can be made between zero and non-zero masses of particles of the same spin. A spin-one and zero mass "photon" has but *two* helicities $\hbar$ and $-\hbar$ while a "massive photon" would also have a spin of one unit but *three* helicities $+\hbar$, $-\hbar$ and 0.

clockwise along their direction of motion (*right-handed*) or anticlockwise (*left-handed*). This "spinning" motion with its correct magnitude – as well as the correct "handedness" – is something which emerges from the Dirac equation in a natural way – it was not put into it by hand – and this is one of the equation's great triumphs.

The second point about the Dirac equation is that it predicts that each particle has an antiparticle – with the same mass and intrinsic spin but opposite electric charge* (if any). Furthermore, a charged particle and its antiparticle can annihilate each other, the surplus energy going into the production of photons (γ's).

As I said earlier, Dirac came to his equation by attempting to unify quantum mechanics and special relativity. He found that as well as describing positive energy particles, his equation described "negative" energy objects also.

Now no one had seen these negative energy objects. They would act like proverbial mules. If you

* Conventionally, an antiparticle is written with a bar. For example, if p denotes a proton, $\bar{p}$ stands for an antiproton. In any elementary particle reaction, one can transfer particles from one side of the equation to the other, provided they are replaced by antiparticles on the other side. Thus $n + \nu \leftrightarrow p + e$ is equivalent to $n + \bar{e} \rightarrow p + \bar{\nu}$, $e + \bar{\nu} \rightarrow \bar{p} + n$, $\bar{n} + p \rightarrow \bar{e} + \nu$ or to $n \rightarrow p + e + \bar{\nu}$.

pulled them forward they would move backwards! And in quantum theory one could not get rid of these negative energy objects by fiat. One had to find a new interpretation for these.

There is the story that Dirac got the idea which enabled him to resolve this difficulty while attending a (Christmas) Archimedean's problem-drive at St. John's, where he was a student. The story is probably apochryphal, but I will tell it all the same.

All those who participated in the problem-drive were posed 'quickies' – problems which they had to solve in a few seconds. Dirac was given the following problem:

Three fishermen go fishing in the middle of a stormy night. They catch some fish, then land on a desert island and go to sleep. Later, one of them wakes up and he says to himself 'I'll take my one-third of the fish and I'll go away.' So he divides all fish into three equal parts. He finds one fish extra. He throws the extra fish into the sea, takes his one third, and goes. Next, the second fisherman gets up. He doesn't know that the first one has departed. He too divides the remaining catch into three equal parts. He also finds one fish extra, throws it into the sea and leaves with his one-third share. Finally, the third fisherman gets up. He does not know what the others have done, but also decides to take his third and leave early. He, in his turn, finds one fish extra which he throws into the sea.

The question was, 'What is the minimum number of fish?' Quick as a flash, apparently, Dirac answered 'minus two fish'. His reasoning was: minus two is minus one, minus one, minus one plus one. The plus one fish is the extra fish which is thrown into the sea. You take away the minus one fish – your share. This leaves minus two fish again for the next fisherman to divide and so on. (Clearly, you must add a definitive positive number to minus two in order to arrive at the sensible (positive) number of fish. This I shall leave as a problem for you.)

Reinterpretation of Dirac's equation and the antiparticles

What is the reinterpretation by Dirac so that the negative energy solutions make sense? Dirac's crucial step was to take the lowest possible energy state – the one with *all the negative* energy states filled with electrons – to be the one in which no particles are observed (the so-called "vacuum" state).* He could now interpret any unfilled *hole* in negative energy states as positive – what he called anti-electrons.

Everything about this interpretation is consistent – including the case where an external electric field is present. It is not difficult to see that the *anti-electron*, (i.e. a negative charged hole of negative energy) would describe a "positron" of positive electric charge and positive energy in the reinterpretation.

Thus it was that Dirac was able to predict the existence of a new particle, the positron – the antiparticle of the electron. This was actually discovered a few years after the Dirac postulation of its existence. This was a triumph, but the still greater experimental triumph was the production of antiprotons by Segrè and Chamberlain in 1956 and the

* This assumes that only one particle can go into each distinct energy state. This assumption, which is true both of helicity $\hbar/2$ and $-\hbar/2$ objects independently, is part of what Dirac and Fermi had earlier postulated for such objects. (In contrast to these spin-1/2 objects are the integer-spin particles, which had been considered earlier still by Bose (and following him by Einstein). Such integer-spin objects are gregarious and tend to congregate into the same energy state, given half a chance).

The half integer-spin $\hbar/2$, $3\hbar/2$, $5\hbar/2$, ... particles are called "fermions" while those with integer-spins 0, $\hbar$, $2\hbar$, ... are called "bosons". The fermions are individualists while the bosons exhibit the "buffalo-herd" collectivist instinct of congregating together.

later production of (composite) antideuterons by Zichichi *et al.* in 1965.*

The equation of Dirac and its successful reinterpretation is one of the greatest triumphs of twentieth century physics. It led to an unbounded adulation of Dirac illustrated by my next story (something I witnessed myself at the 1961 Solvay conference). The story involves Feynman – the greatest physicist of my generation – who, I believe, was the first Dirac lecturer.†

Those of you who have attended the old Solvay conferences will recall that on these occasions one sat along long tables that were arranged as if one was going to pray. Like a Quaker gathering, there was no fixed agenda; the expectation – seldom belied – was that someone would be moved to start

* With his emphasis on half-integer spins, Dirac not only started a revolution in physics but also one in mathematics, through the so-called spinorial calculus. His intuitive mathematical acumen had earlier been established by his invention of the delta function which, in the hands of mathematicians like Laurent Schwarz, has become a whole new discipline of distribution theory. Dirac was a genius both in mathematics and in physics.

† Richard Feynman's 1986 Dirac Memorial Lecture, the first in this series, was published in 1987 in *Elementary Particles and the Laws of Physics* (Cambridge University Press, 1987). The second Dirac Memorial Lecture, presented by my good colleague Steven Weinberg, is also included in that same volume.

the discussion off spontaneously. At the 1961 Solvay Conference I was sitting next to Dirac, waiting for the session to start, when Feynman came and sat down opposite. Feynman extended his hand and said, "I'm Feynman." Dirac extended his hand and said, "I'm Dirac." (Apparently, this was the first time they formally introduced themselves, at least during that conference.) There was silence, which from Feynman was rather remarkable. Then Feynman, like a schoolboy in the presence of a master, said to Dirac: "It must have felt good to have invented that equation" and Dirac replied, "But that was a long time ago." There was silence again. To break this, Dirac of all people asked Feynman: "What are you working on yourself?" and Feynman answered "Meson theories." Dirac said "Are you trying to invent a similar equation?" Feynman said "That would be very difficult." And Dirac said in an anxious voice, "But one must try." At that point the conversation finished because the meeting had started.

So this is one of Dirac's great contributions to the theory of elementary particles: has famous equation which can describe the spins and helicities of the elementary entities like the electrons, the quarks and their antiparticles and also, of the free proton and the free neutron (and *their* antiparticles).

Unification of fundamental forces

Before I talk about the nuclear forces (through which these particles interact), let me tell you how *I* learned about the fundamental forces of physics in 1934.

The Jhang School

I still remember the school at Jhang in Pakistan (Jhang is my birthplace). Our teacher spoke of gravitational force. Of course, gravity was well-known and Newton's name had penetrated even to a place like Jhang. Our teacher then went on to speak of magnetism; he showed us a magnet. Then he said, "Electricity! Ah, that is a force which does not live in Jhang, it lives only in the capital city of this province, Lahore 100 miles East." (And he was right. Electricity came to Jhang *five* years later.) And the nuclear force? "That was a force which lived only in Europe. It did not live in India (or Pakistan) and we were not to worry about it." But I still remember he was very keen to tell us about one more force – the capillary force.* I always wondered why he was so insistent on calling the capillary force "a fundamental force of nature." I think I know

* This particular lesson has stuck in my mind because of the extra-ordinary heaviness of the Arabic word for capillarity. 'Kesh-e-shay anaabeeb-e-shaari'

now the reason. He was teaching us the force laws according to Avicenna.* Avicenna was not only a great physicist of distinction but also a great physician. Surely for a physician there is no force more important than the one which makes blood rise in the smaller capillaries. He (and my teacher) regarded it as one of the fundamental forces of nature– though *we* do not think so today.

The nuclear forces (*continued*)

For our purposes, the forces of central importance (besides gravity and electromagnetism) are the two nuclear forces – the "strong" and the "weak". To describe some basic aspects of these, I will concentrate on four objects: the proton p^+, the neutron n^0, the electron e^- and the neutrino ν_e^0, where the superscripts (+, −, 0) give the electric charges on these objects†.

* Avicenna was a contemporary of Al-Biruni. He was a native of central Asia, who wrote both in Arabic as well as Persian. The Russians claim him as "The First Great Soviet Physicist" because his birth-place is now part of Soviet Central Asia. (At the Moscow University, Avicenna's statue occupies the leading place of honour.) His Qanun ("Canon of Medical Law") was taught in Europe till the 17th century.

† The leptonic doublet (ν_e^0, e^-) is supposedly elementary, while the doublet (p^+, n^0) consists of composite objects and is made up of quarks (u, d).

The Strong and the weak nuclear forces

As I said before, the "strong" nuclear force acts only between the members of the (p^+, n^0) doublet. The "strong" nuclear force has a range of around 10^{-13} cm. This is the force which is responsible for nuclear fission and for nuclear fusion (the energy source of the stars).

The weak nuclear force

In contrast, the "weak" nuclear force is *nearly* universal. It can act between members of the leptonic doublet

$$(\nu_e^0, e^-) \leftrightarrow (\nu_e^0, e^-)$$

or between $\quad (p^+, n^0) \leftrightarrow (\nu_e^0, e^-)$

or between $\quad (p^+, n^0) \leftrightarrow (p^+, n^0)$

The weak nuclear force was first discovered by Madame Curie as the force responsible for the so-called β radioactivity. It plays a crucial role in the energy production by the sun. *The weak nuclear force is universal but not as universal as gravity.* In 1957, it was discovered that the "weak" nuclear force, to our knowledge, acts only between left-handed particles.* *There is no "weak" force between*

* A left-handed particle is a spin-1/2 object, spinning like a rotating top in an anticlockwise direction relative to its direction of motion.

right-handed electrons, protons or neutrons while right-handed neutrinos may not even exist.

This predilection for left-handed force-laws clearly violates the symmetry of mirror reflection. (A mirror reflects a right to a left hand.) When one reflects a (left-handed) neutrino in the mirror, one may see nothing.

The "weak" nuclear force is called weak because (at comparable distances) it has 10^{-5} times the strength of the electromagnetic force. The range of the weak force is no more than 10^{-16} cm (which is shorter, by a factor of 1000, than the range of the strong nuclear force).

Summary

To summarise, we have considered particles of spin $\hbar/2$. These fall into two categories – the six light particles whose left-handed components exhibit the weak nuclear force of a range of the order of 10^{-16} cm, and the strongly interacting quarks which make up the protons and the neutrons. These interact through a strong nuclear force also of short range (of the order of 10^{-13} cm). In addition to the two (weak and strong) nuclear forces, there is of course the electromagnetic force between the electrically charged protons and electrons plus the universal force of gravity, making in all the four fundamental forces of nature.

4. THE UNIFICATION OF THE WEAK NUCLEAR FORCE WITH ELECTROMAGNETISM

Forces produced by exchanges of messengers

The crucial idea that permitted our generation to unify electromagnetism and the weak nuclear forces (rather than electromagnetism and gravitation, as Faraday and Einstein wanted) was that both these forces (electromagnetism and the "weak" nuclear) have spin-one messengers and are "gauge" forces. (I shall define "gauge" forces presently.)

In quantum theory, all forces – whether of gauge or nongauge variety – are produced by an exchange of particles, which I shall call "*messengers*". These "messengers"* must always have integer intrinsic spins (0, $1\hbar$, $2\hbar$, etc.) as contrasted with the "source" particles of matter like electrons, neutrinos or protons and neutrons which are described by Dirac's equation and have intrinsic spins of $\hbar/2$ each.

The fundamental property that characterises

* I chose the rarer name "messengers" rather than the commonly used word "mediators" in deference to nomenclature for RNA. (I believe the double-helix ideas were discovered in the Austin wing of what in my days was the "new" Cavendish Laboratory. Application of these ideas to hemoglobins were elaborated essentially in the converted sheds next-door! Not many know about this but I was wondering if some plaque had been put up to say that Max Perutz and his coworkers (including Kendrew, Watson and Crick) were located in these huts during the 1950s and the 1960s).

gauge forces (as opposed to nongauge forces) is that they are produced by an exchange of "messengers" of *spin-one.**

The gauge forces and "messengers" of spin-one

The prototype for all *gauge forces* is electromagnetism. Here the "messenger" – of spin-one – is the photon, γ – the quantum of light. To picture the force between "source" objects like a proton (p^+) and an electron (e^-), each carrying an electric charge of unit magnitude (but with opposite signs) we may imagine a proton coming along and emitting a photon, after deceleration (change of direction) in accordance with Maxwell's ideas (see Fig. 4*a*). The photon is next absorbed by the electron (e^-), which accelerates (Fig. 4*b*), again in accordance with Maxwell's ideas. If the photon is not actually seen, the net effect of this exchange process for any observer is the change of momentum *between* the

* The gauge principle embodies universal behavior which connects the strength of gauge forces with the notion of charge. Thus for example, the strength of the gauge force between the photon and helium nucleus is twice that between the photon and hydrogen nucleus because the electric charge on the helium nucleus is twice the charge on hydrogen.

proton and the electron. This is what we call the *force of electromagnetism* in operation (Fig. 4*c*).

One can generalise these ideas to *multiplets* of "gauge messengers". One finds that "*gauge messengers*" *of equal masses** can occur grouped together only in multiplets of certain size. For example, the gauge multiplets should consist of *singlets* (1), *triplets* (3), *or octets* (8), etc. ..., but no numbers in between.† *In particular, there are no doublets amongst the possible multiplets of gauge messengers. This is an important remark (not true for non-gauge messengers) which played a crucial role in the gauge*

* The range of the force produced by a "messenger" depends inversely on the rest-mass of the messenger. To produce a weak nuclear force of range 10^{-16} cm the messenger ought to have a rest-mass of the order of 87 proton masses. The electromagnetic force is long range because its messenger (the photon) has a rest-mass of zero.

† For generalised gauge forces, one needs to bring in the mathematical theory of Lie symmetrics, which correspond to rotations in the "internal" space.

Electromagnetism is based on the simplest of all "Lie" symmetries U(1). By using other Lie symmetries, one may obtain other, more elaborate, gauge theories. Thus, instead of a single gauge messenger (e.g. the photon of electromagnetism), there could be several gauge messengers of equal mass placed together in a multiplet. The size of the multiplet depends on the size of the "regular" representation of the "Lie" symmetry concerned. There can only be one messenger for U(1): three for SU(2): eight for SU(3) etc. A further consequence of this is that all gauge particles must have zero masses. (At this point I must apologise to my good friend, Sir Harry Hinsley.)

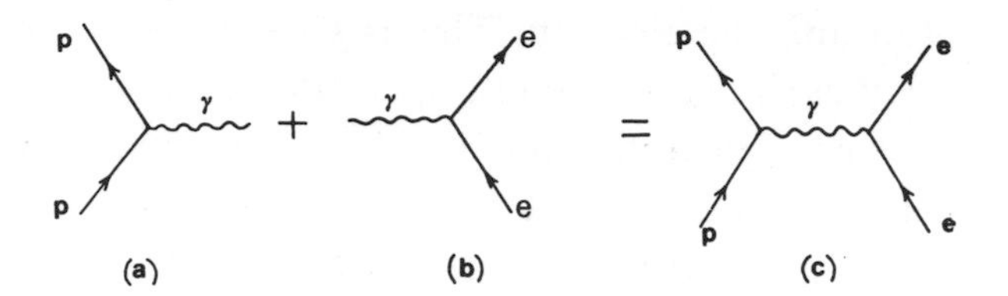

Fig. 4 The electromagnetic force between a proton (p) and an electron (e).

identification and unification of the weak and electromagnetic forces.

The unified electroweak force

When John Ward and I first started to look at the problem of unification in the 1950s, the conventional wisdom about the four fundamental forces was as follows:

(*a*) There was electromagnetism, a gauge force, with a single spin-one messenger the photon. This was well understood as *the gauge force* par excellence.

(*b*) The strong force between the protons and the neutrons was *believed* to have three messengers of *spin-zero* (the pions) from the work of the Japanese physicist, Hideki Yukawa. *This was not a gauge force. A gauge force must be mediated by spin-one particles*; the pionic messengers carried zero spin.

(*c*) The messengers of the weak force were unknown at that time. It was not clear whether they had spin-*zero* or spin-*one* units *if they existed* at all.

(*d*) Finally there was gravity which was thought to require a messenger of spin *two units.*

There did not appear to be much in common between these four forces so far as the messengers and their intrinsic spins were concerned. However, the situation changed as a result of mirror-symmetry experiments suggested by the Chinese physicists Lee and Yang. These experiments established, in 1957, that the *weak "messengers"* (*if they existed*) must definitely have spin-one.

Now the possible existence of *two* of these messengers, W^+, W^-, (*with equal masses*), could be used to understand all the then known weak interactions – and, in particular, Madame Curie's process of β-decay of the neutron, i.e. $n \rightarrow p + e^- + \bar{\nu}$ (see Fig. 5).

If W^+, W^- are indeed spin-one objects, this would incline one to believe that the weak force is a gauge force. There is then something distinctly odd with W^+, W^-. This is because W^+, W^- (just two of them) together make up *only* a doublet (of equal masses). There should, however, be no multiplets of just *two gauge messengers.* There has to be at least one more messenger to go along with W^+ and W^-;

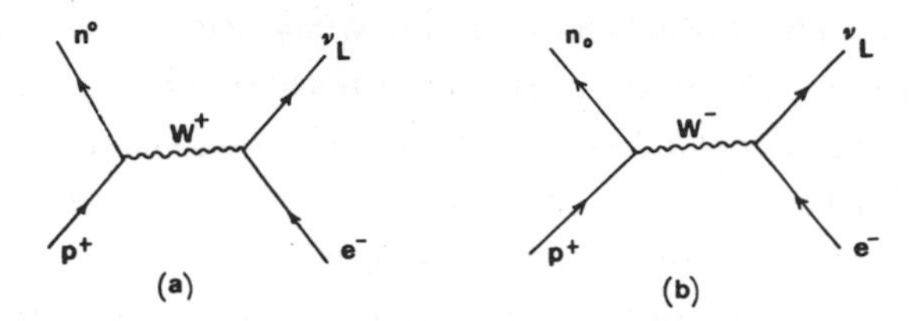

Fig. 5 Two processes equivalent to beta-decay of the neutron: (*a*) W^+, mediated; (*b*) W^- mediated.

together the three ought to make up a gauge triplet.*

It seemed to us – John Ward and myself – only natural (following Schwinger), to suggest that this might be the photon γ, leading to the exciting possibility that one could unify *the weak force* and *electromagnetism* as a single generalised gauge theory. The same idea was proposed independently by Glashow and by Weinberg.

Our first attempt to turn this idea into a credible theory turned out not to be correct: it had a number of problems with it. One major problem was that the weak force was left-handed only, while the electromagnetic force was both left- and right-

* A triplet of gauge messengers would correspond to the "regular" representation of the rotation symmetry in three dimensions. (See pp. 40–41).

handed. For a unified theory, how could one make sense of this discrepancy? The second problem concerned the masses of the gauge particles. If the photon and W^+, W^- make up a gauge triplet, all three had to have zero masses. We knew that the photon had a zero mass but we also knew that W^+, W^- had large masses – large because the weak force has a very short range of the order of 10^{-16} cm. (This meant a mass at least of the order of 87 proton masses.)

To solve the first problem: in theoretical particle physics we have a safe rule due to Marshak. Whenever in doubt, experience has taught us that one should double the number of particles one is dealing with. We proceeded to do precisely this. We doubled the number of neutral messengers by postulating a new object, a "heavy photon" (called the Z^0), which would then make up the desired triplet with W^+, W^-, Z^0 as messengers. The new "messenger" Z^0 would produce a new type of *weak* force between left-handed objects only: for example. $(p + e^-)_L \rightarrow (p + e^-)_L$ or $(p + \nu)_L \rightarrow (p + \nu)_L$, or $(\bar{\nu} + p_Z) \rightarrow (\bar{\nu} + p)_L$, or $(n + e^-)_L \rightarrow (n + e^-)_L$, etc.

This was fine, but with Z^0 treated this way one had lost all trace of unification of electromagnetism and the weak forces. To keep the unification, together with the gauge character of the triplet, we had to use all the four messengers, W^+, W^-, Z^0 and γ.*

Why were we so keen to unify? This was because of Dirac's first requirement that the final theory should be free of inconsistencies and infinities. In order to achieve this – in a renormalizable version of the theory – one needed both Z^0 and γ^0 *mixed* inextricably together. (Any attempt at "unmixing" Z^0 and γ^0 would bring back Dirac's inconsistencies and infinities as was demonstrated by G. t'Hooft and B. W. Lee.)

The next problem we had to face involved the masses of the gauge particles. The photon (γ) must be a particle of zero rest-mass, giving rise to the long-range character of electromagnetism. In contrast, the W^+, W^- and Z^0 needed to be heavy in order to produce the short-range weak forces of range 10^{-16} cm. Where did this mass come from?

Weinberg and I suggested that the answer lay in understanding the environmental factors which

* The final symmetry which was being manifested concretely by these four particles before mixing was SU(2) × U(1), where SU(2) gives rise to the triplet (W^+, W^-, Z^0) of gauge particles while U(1) gives rise to the singlet gamma.

spontaneously break the symmetry. The large masses of the Z and W's must arise through a phase transition (like the ice and water transition which occurs at critical temperature of 0°C). A sheet of water exhibits, for example, a number of symmetries not shared with ice crystals. Below the critical temperature of 0°C, *the symmetry is broken* – above this temperature, the symmetry is restored. *The higher the temperature, the greater the symmetry.*

One of the most well-known examples of (spontaneously) broken symmetry is the theory of ferromagnetism, which was worked out by Werner Heisenberg in 1928. Magnets have north and south poles, which means that they have a preferred orientation in space; they are not symmetric. If a bar-shaped iron magnet is heated, (it loses its magnetic properties and) it becomes symmetric – there is no way of distinguishing one end from the other. When the magnet cools down, however, it spontaneously recovers its magnetization and again acquires a north and south pole. The symmetry, we say, is broken spontaneously or hidden; it is manifest when the bar is red-hot, concealed when the magnet is cold. Technically one calls a state when magnetism is manifest the state of order or the state of spontaneously broken symmetry. We particle physicists had to learn the importance of order* versus symmetry so far as gauge unification of short-range and long-range forces are concerned.

In particle physics, such a phase transition can be brought about by postulating extra particles – the so-called (non-gauge) Higgs bosons† of spin-zero. We proposed that before the phase transition occurs, the W's, the Z's and photon are all *massless*. There should in addition be a Higgs doublet

* Order is the act of choice among possible symmetrical states. The simplest example of symmetry versus order is a circular dining table, in a country where dining manners have not been standardised. The table is laid out, with a napkin and a piece of bread for each guest in a *symmetrical* fashion.

The guests sit down, from the corner of their eyes they glance at their neighbours, trying to decide which napkin to choose – the one symmetrically placed on their right or their left. Suddenly one bold spirit makes his (her) choice and instantly (with the speed of light) an ORDER is established around the table.

In a gauge field theory, order – the state when W^+, W^- and Z^0 acquire masses – comes to be established through the postulation of Higgs particles with specified parameters of their mutual interactions.

† With the Higgs mechanism (together with the mixing of Z^0 and γ^0) one finds one must have W^+ and W^- weighing ≈ 87 proton masses, while Z^0 should weigh ≈ 97 proton masses.

(H^+, H^0) and its antidoublet $(H^-, \bar{H}^0)$, all four particles carrying zero helicity. After the phase transition when the symmetry is spontaneously broken, the photon remains massless but the $W^\pm$'s acquire masses by incorporating into themselves the charged Higgs $H^\pm$ particles, while Z^0 acquires mass through the incorporation of a part of the neutral Higgs $(H^0 + \bar{H}^0)$ – leaving the other part $(H^0 - \bar{H}^0)$ of the Higgs with a specified value as a *live* object to be experimentally discovered. It is this which gives precise numbers like 87 for $W^\pm$ masses, and 97 for Z° mass. (The emergence of this non-zero value of masses of W and Z is the essence of *spontaneous* symmetry breaking.)*

The total number of helicity states in the W, Z and Higgs system remains the same before and after the phase transition. Before the phase transition, $W^\pm$ and Z^0 are massless with just two helicities $(\pm\hbar)$, while the four Higgs objects correspond to four zero helicity states, so that the total is $6 + 4 = 10$. After the phase transition (at a lower temperature), the helicities are three each for the

* Peter Higgs himself is a very much live Scottish physicist who suggested during 1963 the mechanism for spontaneous symmetry-breaking, giving masses and the additional zero-helicity to otherwise massless gauge messengers, though he did not apply these ideas to the case of electroweak theory. (Unfortunately, unlike the case of $W^\pm$ and Z^0, one cannot predict the mass of the live Higgs particle.)

massive objects W and Z ($\hbar$, $-\hbar$, 0) plus one helicity for the one live neutral Higgs remaining (i.e. $3 \times 3 + 1 = 10$ once again).

As for the critical temperature, one finds that this electroweak phase transition must occur when the universe had a temperature of around 300 proton masses.* This temperature of around 300 proton masses was a consequence of the desire to unify electromagnetism with weak forces. From Friedmann's work, such indeed was the temperature of the Universe some 10^{-12} seconds after the Hot Bang.

Before the phase transition took place, (that is, when the temperature was higher than 300 proton masses) there was just a *single electroweak force.* Immediately afterwards it split into two distinct forces, *electromagnetism* and the *weak nuclear* – with the W^+, W^-, Z^0 acquiring masses.

Let me summarise. The sequence of ideas we had suggested went something like this: One starts with messengers, W^+, W^- which must have spin-one from the parity violation experiments which were

* To specify temperatures, I am using a set of units measuring temperatures in terms of masses. If one was using degrees, then one proton mass would be the equivalent of 10^{13} °C. Thus, in degrees, the critical temperature (of 300 proton masses) is equal to some 3×10^{15} °C. The symmetry is restored and $W^{\pm}$ and Z^0 become massless at temperatures higher, i.e. at times earlier than 10^{-12} seconds.

suggested by Yang and Lee. If these are gauge messengers, one must postulate a third photon-like messenger Z^0. The Z^0 produces new types of weak forces.

One must mix in the old photon γ together with the new photon-like Z^0, in order to save the theory from infinities and inconsistencies in the sense of Dirac. All the four particles W^+, W^-, Z^0 as well as the photon at this stage would have zero rest-masses.

The masses of the heavy objects (W^+, W^- and Z^0) come about because of a phase transition which occurs at a temperature of around 300 proton masses in the early universe. To bring about this phase transition, we need a doublet and an anti-doublet of (nongauge) Higgs spin-zero particles to start with.

I shall go quickly over the subsequent experimental history of the subject.

(*a*) We had predicted a new type of messenger particle, the uncharged Z^0, which should produce a new type of weak force, and where, because of the electric neutrality of the exchanged Z^0, the *final* and *initial* particles must be the same.* Examples are

* This is not the case for the exchanges of W^+, W^- where, for example, the initial particles in the reaction $n + e^+ \rightarrow p + \bar{\nu}$ are not identical with the final set though the total net charge is the same (i.e. one unit) before and after.

shown in Fig. 6. Such events were discovered for the first time at the European Laboratory for Particle Physics (CERN) during 1973 using the Gargamelle bubble chamber – a gift from the French Government.

(*b*) The presence of this new force (due to the exchange of Z^0) could also be seen in the interference between photon-(γ^0)-mediated and Z^0-mediated interaction of an electron and a deuteron (d^+)*, see Fig. 7*a* and 7*b*. Normally, with γ^0 exchanges alone, the force arising would have been considered an electromagnetic force with both left- and right-handed electrons being scattered in 50:50 ratios. But, with the interposition of Z^0 (with new left-handed weak forces $(p + e^-)_L \rightarrow (p + e^-)_L$ *and* $(n + e^-)_L \rightarrow n + e^-)_L$) the 50:50 balance would be upset. The asymmetry of left versus right scattered electrons was looked for at the Stanford Linear

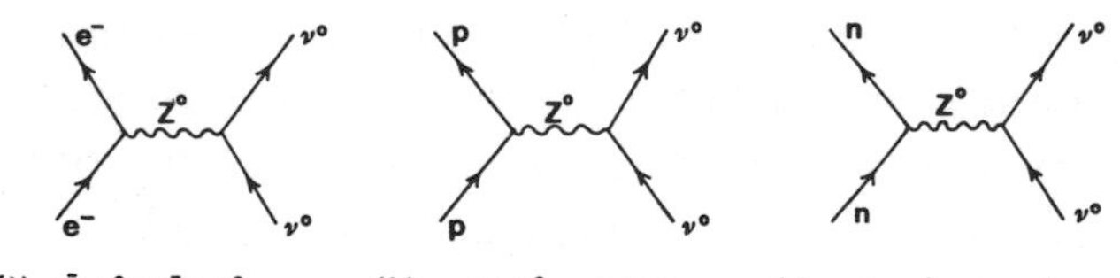

Fig. 6 Examples of "neutral current" processes mediated by the Z^0 boson.

* The deuteron is a composite of a proton and a neutron.

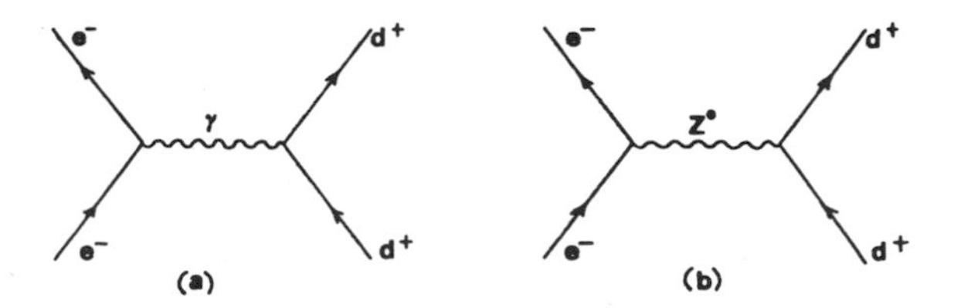

Fig. 7 The reaction between an electron and a deuteron can be mediated by either a photon or the Z^0 boson. Interference between the two processes, in accord with the predictions of the new electroweak theory, provided indirect evidence for the theory's validity. (The deuteron is made up of one proton and one neutron.)

Accelerator Center (SLAC) in 1978, and found – precisely in accord with predictions.

Note that both these two experimental tests at CERN and at SLAC were "indirect" tests. One does not actually produce the Z^0 particle into the open air. (There was not enough energy to make live Z^0's).

We needed direct tests where live Z^0's were to be produced. As I said earlier, the masses of the Z^0 and of $W^{\pm}$ were predicted to be of the order of 97 and 87 proton masses while their intrinsic spins must be unity. These particles, with the mass values as predicted (correct up to 1%), were actually produced at the CERN laboratory during 1983. Also the spins of the W^+, W^- and Z^0 were found to be unity.

To produce W^+, W^- and Z^0, one had to modify the CERN accelerator, SpS, to Sp$\bar{p}$S (Super proton-anti-proton Synchrotron), in order to produce in the small interaction zone the same conditions as existed in the early universe. Specifically, a new antiproton beam was created using what has now come to be called an antiproton accumulator. The antiproton $\bar{p}$ annihilates the p, producing a total energy in excess of what was needed for producing W^+, W^- and Z^0. This epic experiment by Van der Meer who engineered the $\bar{p}$ beam and Carlo Rubbia who engineered the whole enterprise won the first Nobel prize for CERN (European Organization for Nuclear Research) in 1984.

Experiments are now planned at the new SLAC (Stanford Linear Accelerator Center) and the forthcoming LEP accelerators at CERN to measure the Z^0 mass to one-twentieth of one percent. If any deviations from the electroweak theory are found (and one hopes for such deviations for the future of the subject) the theory would need to be *extended* – a consummation profoundly to be desired.

Why must the theory be extended? This is mainly because the theory at this point contains twenty-one parameters corresponding to the masses of the

Fig. 8 (*a*) The theorists (from left to right) Sheldon Glashow, Abdus Salam, and Steven Weinberg were awarded the 1979 Nobel Prize for Physics for their theoretical contribution to the unification of the weak and electromagnetic forces. (*b*) Carlo Rubbia (left) and Simon van der Meer (right) were awarded the 1984 Nobel Prize for Physics for their contribution to the experimental verification of the existence of the W and Z bosons which were first produced and detected at CERN in 1983.

quarks and leptons. All these essentially are "free" parameters which have to be specified by having recourse to experimental information. This is not satisfactory for a *fundamental* theory which should naturally discover all these numbers in terms of just one fundamental mass (say the Planck mass),* – the only free parameter that specifies the size of our universe and all that is contained in it.

Z^0 mediation and "handedness" of biological molecules

There is an interesting consequence of the existence of the new weak left-handed interactions (the Z^0 forces) and this brings me back to the cycle sheds outside. It is well-known that a number of biological molecules can exist in two distinct forms; these have the same chemical formulae but are mirror images of each other. An example is provided by the L-alanine and D-alanine molecules shown here (see Fig. 9).

In chemical synthesis in the laboratory, left-handed and right-handed molecules are made in equal amounts; it's a fifty-fifty mixture. But in nature (in general), only left-handed amino acids and right-handed sugars are found. Understanding

* See the footnote on p. 21.

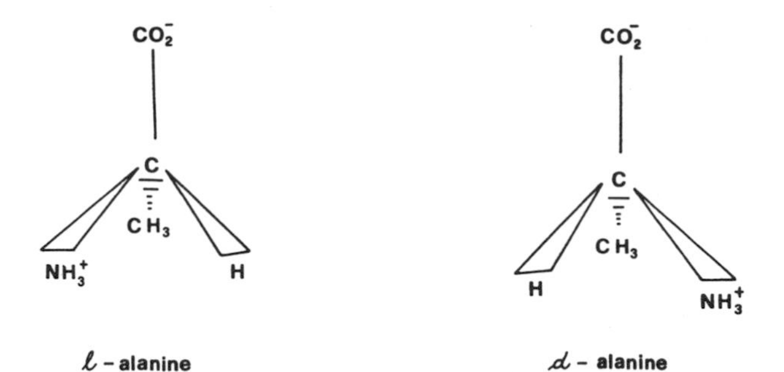

Fig. 9 Representations of the mirror image molecules L-alanine (left handed) and D-alanine (right handed).

this handedness of nature is of vital importance. For example, the drug “thalidomide” (which caused birth defects) was made in the laboratory and administered as an equal mixture of the left- and right-handed forms. Later is was found that it is the left-handed form which gives rise to birth defects. A further example concerns penicillin. Bacteria, exceptionally, utilise right-handed D-amino acids in the construction of their cell walls. Penicillin itself contains a group of right-handed amino acids and interferes with the synthesis of the bacterial cell walls, a process which is unique to bacteria and which does not occur in the mammalian host.

Dr. S. Mason, his collaborators at King’s College, London, and others, have provided a possible explanation for the asymmetry of biological molecules in terms of the broken mirror symmetry of the weak forces.* What they did was to consider the left-handed Z^0 force, *together with* the well-known photon mediated force of electromagnetism. (This last was what had previously been used to determine the energies of the molecules and their structures). They found (that with the Z^0 force taken into account) the left-handed amino acids (and right-handed sugars) are more stable than the oppositely handed molecules. According to these calculations, there should be an excess of about one part in 10^{16} of the more stable variety initially.

Next, one has to use some form of "catastrophe theory".† Once a preference for one mirror state is established – even if it is a *very small preference* – then the long biological times over which biological matter has lived is supposed to do the rest, explaining why none of the less stable molecules are observed in nature today.

There is a growing confidence today that the electroweak force is the true "force of life" and that the Lord created the Z^0 particle to provide handedness for the molecules of life.

* *New Scientist*, 19 January 1984.

† This argument needs further refining.

5. The strong nuclear force as a gauge force and the standard model

The strong nuclear force and the gluons as gauge particles

Together with the "electroweak" force, comprising the weak nuclear forces and electromagnetism, there have been parallel developments in identifying the *gauge* aspects of the strong nuclear force, also mediated by messengers of spin-one – the so-called gluons (so that one has come to believe more and more in the validity of the gauge ideas). These developments started in the early 1970s and culminated in the "indirect" findings of gluons at the DESY Laboratory at Hamburg in 1979.

The story starts with three doublets of quark flavours (u, d), (c, s) and (t, b). It turns out that there are not just six but eighteen distinct quarks, distinguished from each other by COLOUR. Each quark comes in three colours which have once again been given whimsical names like Red (R), Yellow (Y) and Blue (B). A postulated symmetry between these three colours* would give rise to *eight* gauge particles – the so-called gluons. There is no Higgs particle for colour; the symmetry remains un-

* The symmetry would be SU(3) for three colours, with eight gauge objects (in its "regular" representation of spin-one).

broken, and the messenger particles, the gluons, remain massless. Detailed tests have supported the idea of the strong nuclear force being produced by such messenger gluons.*

The Standard Model†

At present then we can describe all elementary entities plus all forces of nature (except for gravity) as shown in Tables 1 and 2.

Table 1. *Elementary entities of matter.*

Quarks

Two flavours, three generations, two helicities (left and right, L, R). Three colours, Y, R, B stand for red, yellow, and blue

$$\left.\begin{matrix} u_R, d_R \\ u_Y, d_Y \\ u_B, d_B \end{matrix}\right|_{L,R} \qquad \left.\begin{matrix} c_R, s_R \\ c_Y, s_Y \\ c_B, s_B \end{matrix}\right|_{L,R} \qquad \left.\begin{matrix} t_R, b_R \\ t_Y, b_Y \\ t_B, b_B \end{matrix}\right|_{L,R}$$

Leptons

$$\begin{matrix} (\nu^0_{eL}, e^-_L) & (\nu^0_{\mu L}, \mu^-_L) & (\nu^0_{\tau L}, \tau^-_L) \\ (e^-_R) & (\mu^-_R) & (\tau^-_R) \end{matrix}$$

* We believe that the gluons as well as the quarks – in fact all coloured objects – are permanently confined within protons and neutrons. Permanent "confinement", an empirical finding, is a notion for which there is no good theoretical explanation yet.

† The Standard Model corresponds to the *gauge* symmetry SU(3) × SU(2) × U(1).

Table 2. *Fundamental force messengers and the Higgs particle*

Electroweak gauge messengers, W^+, W^-, Z^0, γ^0.

W^+, W^-, Z^0 are massive, with three helicities $\pm\hbar$, 0 each while γ^0 has zero rest-mass (and therefore just two helicities $\pm\hbar$).

Strong nuclear gauge messengers

Eight (electrically) neutral massless gluons of spin $\hbar$ (helicities $\pm\hbar$). There is no spontaneous symmetry breaking here.

The Higgs particle

One live neutral Higgs object (of zero helicity) left over as a consequence of the (spontaneously) broken electroweak symmetry.[a]

[a] "The discoveries of recent decades in particle physics have led us to place great emphasis on the concept of broken symmetry. The development of the universe from its earliest beginnings is regarded as a succession of symmetry breakings. As it emerges from the moment of creation in the Big Bang, the universe is completely symmetrical and featureless. As it cools to lower and lower temperatures, it breaks one symmetry after another, allowing more and more diversity of structure to come into existence. The phenomenon of life also fits naturally into this picture. Life too is a symmetry-breaking." (F. J. Dyson, *Infinite in All Directions*, Harper and Row, Cornelia and Michael Bessie books, 1988).

The objects shown in Tables 1 and 2 comprise the 'Standard Model' of elementary particles as of 1988.

The Higgs particle of the Standard Model has not yet been produced in the laboratory (nor have the top quarks) but all the rest of the particles are known to exist directly.

6. BEYOND THE STANDARD MODEL

What are the ideas beyond the Standard Model which may extend it?

The prime idea relates to what is called supersymmetry. This is a symmetry that is postulated to exist between spin-0 and spin-1/2, or spin-1 and spin-1/2, or spin-2 and spin-3/2.

As stated before, particles with integer intrinsic spins (0, $1\hbar$, $2\hbar$, $3\hbar$, ... etc.) are called *bosons*, and those with half-integer intrinsic spins $\hbar/2$, $3\hbar/2$ etc. are called *fermions*. Bosons and fermions are thought to be very different objects – the fermions are individualists, the bosons are collectivist — but supersymmetry can relate them in a precise way. *At one stroke, supersymmetry would double the number of objects so far as the Standard Model is concerned.* Thus, in addition to W^+, W^-, Z^0 and γ, there would exist their spin-half partners (called Winos, Zinos and photinos) plus spin-zero friends of quarks and leptons, the so-called squarks and sleptons (unspeakable names).

Supersymmetry is a beautiful notion. It diminishes considerably Dirac's infinities and inconsistencies, though it does not eliminate them completely. It is, however, a symmetry which has received no experimental support as yet (we believe this is because of the low energy of the accelerators so far available).

I shall tell here a story about supersymmetry and Dirac. Dirac always trusted in beauty. For him the beauty of a theory determined whether or not it should be accepted, even against any *presently contrary* experimental evidence. As an example, he had always stressed the beauty of the ideas which went into the 1905 Special Theory of Relativity, particularly of mass depending on velocity which was a cornerstone of the Special Relativity Theory of Einstein. At one time, Dirac recalled, the experimental evidence was against this theory. Dirac maintained that anybody with any sense would have rejected the *experiments as being incorrect* because they went against a beautiful and fundamental theory like the Special Theory. So it proved after the experiments had been refined.

I remember lecturing on supersymmetry in Dirac's presence at the Miami Conference in 1974. Dirac was sitting at the back of the room and had as usual said nothing throughout the lecture. I went up to him and said "Professor Dirac, don't you think this is a beautiful theory? Doesn't it satisfy your criterion for being correct?" He conceded that "it was indeed a beautiful theory but if supersymmetry was a true symmetry of nature, these *new* fermions and bosons would have been found long ago!" I was amazed because it seemed to go against his own dictum of the primacy of beauty. (Intuitive-ly, he may prove to have been right after all; one never knows in our subject.)

Other ideas beyond the Standard Model relate to:

(*a*) Elementarity of quarks and leptons. Is there a next layer of more elementary entities in place of those objects? This is quite on the cards because three *similar* generations is *two* too many.

(*b*) Right-handed weak forces may exist and may show themselves at a higher energy (this may necessitate new $Z^{0\prime}$ to exist and restore the left-right symmetry at a very high temperature, above the masses of Z^0 and $Z^{0\prime}$).

The grand unification of the electroweak force and the strong force

There is the further stage of unification after the unification implied by the Standard Model. Here we try to unite the strong nuclear force with the "electroweak" force. The ideas follow directly on from the electroweak unification – with new gauge messengers and new non-gauge Higgs particles.

This theory is commonly known as *Grand Unified Theory.**

One prediction of such theories is that the proton must eventually decay. This is desirable so far as early cosmology is concerned. If the proton decay does take place (albeit rather slowly with a predicted lifetime of 10^{32} years),† together with time symmetry violation, one would obtain an understanding of why *antiprotons are so scarce* in the universe.

There is no conclusive evidence yet for proton decay with experiments in deep caves on earth. Future experiments may have to be performed on the moon to obviate problems with neutrino backgrounds on earth.

Another prediction of the Grand Unified Theories refers to Dirac's monopoles which I shall discuss later.

* The theory was proposed during 1972 (and in a different version in 1974). Jogesh Pati and I called the new gauge force the "*electronuclear force*". This would consist of electromagnetism plus the weak nuclear force plus the strong nuclear force. A similar combination of forces has been christened the "*grand unified force*" — a "grand" name due to Sheldon Glashow and Howard Georgi. This name, in our view, would have been more appropriate to designate the final theory (The Theory of Everything) where gravity was also unified with the electronuclear forces. However....

† Compare this to the present age of the universe – a paltry matter of fifteen billion years.

(a)
(b)

Fig. 10 Apparatus used in modern particle physics. (*a*) The Gargamelle detector at CERN. (*b*) A view of the LEP ring, which will search for new particles after starting operation on 15 July 1989.

Role of accelerators in particle physics

I mentioned before the role of accelerator technology in the verification of the unified electroweak theory. Since I started research in particle physics, the size and cost of the experiments has increased enormously. This has brought us to the stage where, if we are not careful, there may not be the possibility of testing our theories any further.

In the days of the Old Cavendish here in Cambridge, the style of the experimental work was "string and sealing wax". No more now. At CERN, the $Sp\bar{p}S$ ring is six kilometres in circumference, while the ring for LEP (*L*arge *E*lectron *P*ositron Collider to be commissioned in July 1989) will be 27 km (about 17 miles long). The SSC (the *S*uperconducting *S*uper *C*ollider), may have a circumference of about 93 km (58 miles)!

New accelerators

As to the future, Table 3 lists details of some of the accelerators that are being built or planned.

Firstly, there are four accelerators which will soon (before 1999) be commissioned and which one will hear a lot more about in the next few years. These accelerators may or may not discover the Higgs particles, supersymmetry, right-handed *weak*

Table 3. *The major present and future accelerators.*

Year	Accelerator	Centre of mass energy (in proton masses)	Luminosity (cm^{-2} sec^{-1})	Locality
1987	S$p\bar{p}$S	900	10^{30}	CERN
1987	Tevatron	2000	10^{30}	FERMILAB
1987	TRISTAN	60(e^+e^-)	8×10^{31}	Japan
1987	SLC	100(e^+e^-)	6×10^{30}	Stanford
1987	Bepc	4(e^+e^-)	5×10^{30}	Beijing
1989	LEP I	100(e^+e^-)	16×10^{30}	CERN
1993	LEP II	200(e^+e^-)	5×10^{31}	CERN
1991	UNK	3000	10^{31}	Serpukhov
1991	HERA(ep)	320	5×10^{31}	Hamburg
?	LHC(pp)	20 000	10^{33}	CERN
?	SSC(pp)	40 000	10^{33}	Texas
?	CLIC (e^+e^-)	4000	10^{33}–10^{34}	CERN
?	VLLP (e^+e^-)	4000	10^{33}–10^{34}	Serpukhov
?	ELOISATRON (pp)	100 000	10^{33}–10^{34}	Sicily

forces, or decide whether quarks are elementary or not. This depends on the useful energy of the accelerator and on the predicted energy threshold for the process to be observed.

Then there are the pp accelerators which are at present gleams in physicists' eyes. There is the large proton–proton collider (LHC LEP III) which may be superposed in the (ē + e) LEP tunnel of 27 kilometres at CERN (European Organization for Nuclear Research), Geneva. There is the SSC, a proton–proton collider proposed to be built by the US Department of Energy in Texas. The funding requested is of the order of five billion dollars needed over the five years of its construction. One will want to compare this with the pointless SDI – Strategic Defence Initiative programme, which has spent 12.7 billion dollars already. There is also the proposal to build the Eloisatron, the largest ever proton-proton collider with circumference of 200 kilometres, in Sicily.

In a different category, there are the linear electron-positron (anti-electron) accelerators: (1) the CLIC, the CERN (e + ē) proposal); (2) the similar VLLP accelerator in the USSR. The electron–positron (e + ē) machines have the great virtue that the entire energy of the colliding particles is available for experimentation whereas for a proton–proton machines–such as the SSC–only around one-sixth to one-third of the total energy of the colliding particles is usefully employed. This is because each proton is made of three quarks, and only two of these, one from each proton, actually interact, so that the extra energy in the noninteracting particles is simply wasted.

Now, *circular* (electron–positron) accelerators are limited by the strength of the bending magnets which must curve the particles into their circular paths. There is now the prospect of using *superconducting* technology which will allow more powerful magnets to be made more cheaply, but a limitation which cannot be circumvented is due to synchrotron radiation emitted by any charged particle as it moves in its *circular* path. The energy needed (according to Maxwell's ideas) swells as the *fourth power* of the energy of the collider. No generator of electric power could keep pace with the demands for energy made by these circular accelerators.

The second variety of accelerators are the *linear* accelerators. Linear accelerators do not suffer from the same problems as bedevil circular accelerators, such as the synchrotron radiation. However, they are limited by the accessible electric field-gradient used to accelerate electrons (or protons). Presently, the best electric fields which can be commissioned are no larger than one gigavolt per metre. A factor of a thousand improvement may take place, perhaps in about twenty years, with beat-wave *laser plasma* accelerator technology. But even so, the ultimate Planck energy of 10^{20} proton masses will need an accelerator which is *ten light years long*! This brings a vision of dynasties of accelerator physicists, riding out along the accelerator, in suitable spacecraft!

Early cosmology

Another area where particle physics has provided us with important inputs is the subject of early cosmology, so much so that early cosmology has become synonymous with particle physics. This is because phase transitions, which separate one era of cosmology from another, are also the mechanisms through which the one ultimate unified force is converted into two (gravity plus electronuclear), into three (electroweak plus strong nuclear plus gravity), and finally into four forces (electromagnetic plus weak plus strong nuclear force plus gravity as the over-all temperature of the universe goes down). The fact that these transitions occur at high temperatures (ranging from 300 proton masses up to 10^{20} proton masses) and the fact that temperatures higher than 10^6 proton masses are unlikely ever to be realised with man-made accelerators, makes the early universe and cosmology attractive to the experimental particle physicist as providing the only laboratory which will, at least indirectly, be able to test our theories (through detection of relics of earlier higher temperature eras which may still survive today).

One can distinguish three eras of cosmology:

(*a*) The most recent era, which started around 10^{12} seconds (10^5 years) after the Hot Bang, with Penzias–Wilson radiation, and continues till today – 10^{18} seconds after the Hot Bang. This is the *Large-Scale Matter Era* in which galaxies and superclusters have evolved. We know the physics of this era but the astrophysics is still misty.

(*b*) The second, the so-called *Electroweak Era*, started with the phase transition corresponding to spontaneous breaking of the electroweak symmetry at a temperature around 300 proton masses – which, according to the calculations of Friedmann, took place around 10^{-12} seconds – and continues up to the emission of the Penzias-Wilson radiation.

(*c*) The third and the earliest era is the *Speculative Era*, which may have extended from 10^{-43} seconds after the Hot Bang and continued up to 10^{-12} seconds. During this long period, string theories in two dimensions (see later) may have given rise to four-dimensional space-time as we know it today.*

Among the problem areas with this era are the problems posed by the cosmological constant and the problem of nonobservation of magnetic monopoles.

The cosmological constant was introduced by Einstein into general relativity in order to provide a repulsive force (in addition to the attractive gravitational force) so as to make the universe static. Then Hubble discovered that the universe was actually expanding. There was no longer any need for the cosmological constant. (Einstein called the introduction of this constant the greatest blunder of his life.)

We believe that in any *natural* theory of microscopic physics, one should have good reasons to set this constant equal to zero. Such a reason has been claimed very recently in the theory of wormholes, which may connect our Universe to another (baby) Universe. This could be part of the Speculative Era 10^{-43} seconds after the Bang.

Also included in this era would be the speculative inflationary phase of the Universe's expansion which would guarantee that the density of magnetic

* The theorist, of course, loves this era because one can speculate to one's heart's content.

monopoles is diluted to the present limits of observation on account of the rapid expansion implied by this inflation. (Dirac had introduced such monopoles earlier to provide complete symmetry between electricity and magnetism. These are now part of any self-respecting Grand Unified Theory.)

Non-accelerator experimentation and cosmology

There are many dilemmas in particle physics and cosmology which need non-accelerator experimentation, like neutrino oscillations. Do ν_e, ν_μ, ν_τ, for example, turn into each other, and over what distances? Does Dark Matter* – invisible electromagnetically to our telescopes, and only showing its presence through weak force or gravitationally, which may, in the modern cosmologist's view, constitute as much as 90% of all matter in the universe–exist at all? Could the ubiquitous neutrinos, after all, be the dark matter?

Unification of gravity with the electronuclear forces

And now we come to the last stage of our quest for unification. Does gravity also unify with the other forces, giving the final realisation of Faraday's and Einstein's dream. Here I come back to Dirac's first

* Non-accelerator experimentation is beginning to use *superconducting* calorimetry where the kinetic energy of the impinging particle would be detected by the amount of heat energy produced by its impact.

criterion. What had stopped this theory even being contemplated so far was the worry regarding the infinities which gravity theory spawns as soon as any higher-order calculations are made with it. (Some, but unfortunately, not all of its infinities were diminished by the use of supergravity ideas.)

This Dirac problem apparently has been cured recently by postulating that the fundamental entities in physics are not point particles but strings which make up loops of *finite* size of Planck length. These strings vibrate in modes like the violin strings and give rise to spins zero, $1\hbar$, $2\hbar$, $3\hbar$, $4\hbar$ ... and, in the supersymmetric versions, in addition, to spins $\hbar/2$, $3\hbar/2$, $5\hbar/2$ ....

Physics would change its paradigm once again with the fundamental entities no longer appearing as point particles but as tiny strings. The mathematics which is needed is the mathematics of 2-dimensional Riemann surfaces; four-dimensional space and time arise as secondary concepts.

There are a number of physical requirements which should be satisfied by a string theory:

(*a*) All source particles (quarks and leptons) plus messengers (like gluons, photons, $W^{\pm}$, Z^{0}) plus Higgs of the Standard Model should be comprised within this framework;

(*b*) It should be a geometrical theory since it must contain Einstein's theory of gravity as part;

(*c*) It should describe gravity without any infinities.

To achieve these three conditions would be a miracle, but this miracle seems to be happening, at least in 10-dimensional space-time where a *unique* superstring theory seems to have emerged, following the work of Green and Schwarz, in the autumn of 1984. The important point is that Einstein's theory of gravity does emerge as a special subunit of the string theory. This justifies the picture of the Glashow snake (see Fig. 11) eating its own tail i.e. microphysics at Planck scale (10^{-33} cm) coming together with macrophysics (10^{28} cm of the Universe's size) described by Einstein's gravity theory. This is the ultimate unification in my view. (Glashow drew the snake some time ago as the final desirable theory, while remaining sceptical about strings himself!)

The desirable space-time which emerges from this unique string theory was, as I said, ten dimensional. A Kaluza–Klein-like compactification of six *space dimensions* would then require a descent to the four dimensions of a realistic space-time. We shall also need to go down from Planck mass of 10^{20} proton masses to 10^{2} proton masses characteristic of W and Z particles.

Unfortunately, the uniqueness in ten dimensions that made the string theories so attractive does not

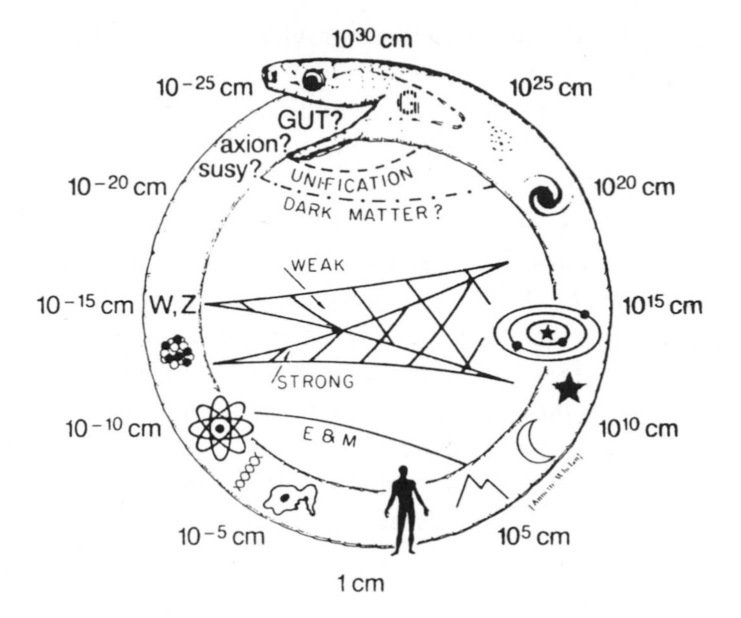

Fig. 11 Glashow's snake eating its own tail. (Reproduced with kind permission from Sheldon Glashow.)

seem to hold when one goes over to four dimensions: a million or more theories (after compactification) appear equally viable. This is one of the theoretical dilemmas that faces string theory at present — in addition to the basic experimental dilemma of building a collider with Planck energy (10 light years long).

Could strings really be the Theory of Everything (TOE) combining all the known source particles, the quarks and the leptons, plus the messengers which we know of, and the Higgs, plus their interactions? If so, would they represent the culmination of one's endeavours to unify the fundamental forces of nature? These are questions which time alone will resolve.

7. Envoi

I would like to leave you with my final thought echoing the words of one of our greatest books:

> Though all the trees on Earth were pens
>   and the Sea was ink,
> Seven Seas after it to replenish it,
> Yet the Words of thy Lord would not be spent;
> Thy Lord is Mighty and All wise.

وَلَوْ اَنَّ مَا فِي الْاَرْضِ مِنْ شَجَرَةٍ اَقْلَامٌ وَّالْبَحْرُ
يَمُدُّهٗ مِنْ بَعْدِهٖ سَبْعَةُ اَبْحُرٍ مَّا نَفِدَتْ كَلِمٰتُ
اللّٰهِ اِنَّ اللّٰهَ عَزِيْزٌ حَكِيْمٌ ﴿٢٧﴾

The Holy Quran (XXXI, 27)

HISTORY UNFOLDING

AN INTRODUCTION TO THE 1968 LECTURES ON THEIR LIVES OF PHYSICS BY W. HEISENBERG AND P. A. M. DIRAC

Abdus Salam

In June 1968, the International Centre for Theoretical Physics in Trieste was privileged to hold an extended symposium on contemporary physics – an event made somewhat unique by its aim to range over and review not just one specialised aspect of modern physics but its entire spectrum.

The most memorable lectures of the symposium were the ones given in an evening series entitled "From a Life of Physics" by some of those to whom we owe the creation of modern physics as we know it. There were six lectures in all: the first five delivered by P. A. M. Dirac, W. Heisenberg, H. A. Bethe, E. Wigner and O. Klein on their own lives of physics; the sixth delivered by E. Lifshitz, commemorating L. Landau's life (who died that year).

The written record of the lectures was published by the International Atomic Energy Agency (IAEA) as a special supplement to the IAEA Bulletin and has not been accessible generally. Since two of the most exciting lectures were those delivered by P. A. M. Dirac and Werner Heisenberg (presided over by Dirac himself) and since both these lectures concern their ideas about the methodology in theoretical physics, it is appropriate that these should also be published in this volume.

In his remarks as chairman of Heisenberg's lecture, Dirac said: "I have the best of reasons for being an admirer of Werner Heisenberg. He and I were young research students at the same time, about the same age, working on the same problem. Heisenberg succeeded where I failed."

This is the height of humility on Dirac's part. However, the mutual admiration of one great physicist for another comes through clearly in these lectures.

If one may make a list of the standing of each one of the great physicists of the twentieth century in the eyes of the others, the list would read something like this – Einstein – Bohr – Heisenberg – Dirac (and Pauli). From Heisenberg's lecture, it is quite clear that Bohr felt happy if he could convince Einstein of the correctness of his ideas. Heisenberg had the same feelings towards Bohr as Bohr had for Einstein, while Dirac considered Heisenberg as the ultimate master. (Pauli, I know, had similar veneration for Dirac.)

I am sure these historic lectures will be as much appreciated as they first were by those present at Trieste in 1968.

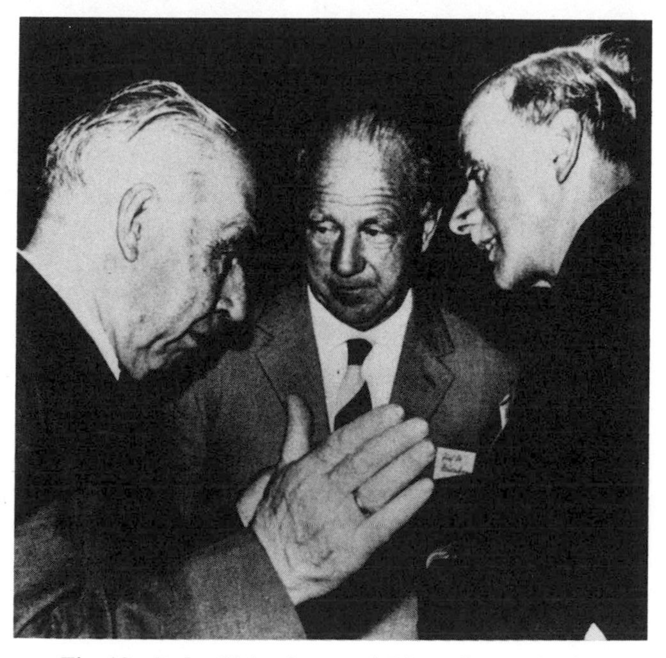

Fig. 12 Bohr, Heisenberg and Dirac discuss the fish that got away.

Fig. 13 Werner Heisenberg.

THEORY, CRITICISM AND A PHILOSOPHY*

Werner Heisenberg

Introductory remarks by Abdus Salam

In 1748 the Shahinshah of Persia, Nadir Shah, invaded India and he marched on to Delhi. He inflicted a severe defeat on the Great Mogul of India. Delhi submitted and the two kings met to negotiate peace. At the conclusion of these negotiations, which included the transfer of the famous Peacock Throne to Iran from Delhi, the Granz Vizier of the defeated Indian King, Asifjah was summoned to present to the two monarchs some wine to pledge the peace. The Vizier was faced with a real dilemma of protocol. The dilemma was this; to whom should he present the first cup of wine? If he presented it first to his own master, the insulted Persian might draw his sword and slice the

* The text of this lecture is reprinted here with kind permission of the International Centre for Theoretical Physics, Trieste.

Vizier's head off. If he presented it to the Persian invader first, his own master might resent it. After a moment of reflection, the Grand Vizier hit on a brilliant solution. He presented a golden tray with two cups on it to his own master and retired saying, "Sire it is not my station to present wine today. Only a King may serve a King". In this spirit I request one Grand Master of our subject, Professor Dirac, to introduce another Grand Master, Professor Werner Heisenberg.

Introductory remarks by P. A. M. Dirac

I have the best of reasons for being an admirer of Werner Heisenberg. He and I were young research students at the same time, about the same age, working on the same problem. Heisenberg succeeded where I failed. There was a large mass of spectroscopic data accumulated at that time and Heisenberg found out the proper way of handling it. In doing so he started the golden age in theoretical physics, and for a few years after that it was easy for any second rate student to do first rate work.

Later I had the great fortune to do some extensive travelling with him.

In Japan, where we were very hospitably entertained, I found how good he is at mountaineering and what a wonderful sense of heights he has. We had to climb a high tower with a platform at the top, surrounded by a stone balustrade. On each of the four corners the stone-work was a little bit higher. Heisenberg climbed up on the balustrade and then on to the stone-work at one of the corners and stood there, entirely unsupported, standing on about six inches square of stone-work. Quite undisturbed by the great height, he just surveyed all the scenery around him. I couldn't help feeling anxious. If a wind had come along then it might have had a tragic result.

THE LECTURE BY WERNER HEISENBERG

First steps in physics

I am indeed very grateful to Dirac for this very nice introduction. I would like to connect the memories I have of the old times of physics with the question about the methods which one uses in theoretical physics. There are so many attitudes which one may have: one can try to formulate phenomenological theories, one can ponder about rigorous mathematical schemes, one can ponder about philosophy and so on and I would like to analyse these various methods in connection with those experiences which I had in this life of physics.

Soon after I entered university, Sommerfeld, who at that time was the Professor of Theoretical Physics at the University of Munich, came to my room and asked me "Well, you are interested in atomic physics, could you try to solve a problem?" I was very interested because I knew nothing about physics, but then he told me that this was quite easy and that it was more or less like solving a crossword puzzle instead of doing rigorous mathematics. The problem was this: He just had received new photographs of spectral lines in the anomalous Zeeman effect. I think that the spectra were taken by the experimental physicist Back in Tubingen, and Sommerfeld told me "There you are, given the lines now you try, according to the theory of Bohr, to calculate or determine the energy levels belonging to these lines in order to describe every line as a difference between two energy levels and then attach quantum numbers to them and then you should be able to reproduce the picture."

One should of course try to find a formula for the

energies as a function of quantum numbers and similar things. This first attempt ended at once with a catastrophe, because I found out that I had to introduce not integer numbers as quantum numbers but half integers, i.e. one half, three halves and so on, and Sommerfeld was terribly shocked to see that. He thought that it was altogether wrong and my friend Wolfgang Pauli, who was also a student in the same seminar, told me "If you introduce half quantum numbers then you will soon introduce quarter of an integer and then a tenth of an integer and finally you will come back to the continuous analysis and we have the classical theory again". After a while there were more of us interested in these problems. There was Pauli, Hönl and others, and it turned out that one really had to introduce half integers as quantum numbers. We had a nice group of young people doing phenomenological physics together, that is, inventing formulae which seem to reproduce the experiments; in this way the formulae of Landé were found and then the multiplet formulae of Sommerfeld and Hönl and so on.

Phenomenological theory

One of these attempts impressed me most and I think I should tell it just to show also the limitations of phenomenological theories.

Sommerfeld told me about an old paper of Voigt in Göttingen, which was written early in 1913, earlier than Bohr's Theory of the Atom. Voigt had given a theory of the anomalous Zeeman effect of the sodium D lines. For doing that he had introduced two coupled linear oscillators arranged such that the two D lines came out, and he could also assume the coupling in such a way that he got the anomalous Zeeman effect; he could even represent the Paschen–Back effect and the intensities and quite generally he could reproduce the experiments extremely well. Sommerfeld asked me again to translate these results into the language of quantum theory, and it turned out that this was easily done. I arrived at rather complicated long formulae for the energy levels and for the intensities, formulae with long square roots containing the square of the magnetic field, the coupling constant, etc., and still the experiments were represented extremely well. I have mentioned this case of a phenomenological description because it fitted so unusually well; but had it anything to do with quantum theory? Six years later we had quantum mechanics at our disposal and Jordan and I tried to calculate the same levels and intensities from quantum mechanics. We got exactly the same formulae as Voigt with the same long square roots and the same intensities. So, in one way you see that phenomenological theories can be extremely successful insofar as they can sometimes give the exact results and consequently agree extremely well with the experiments. Still, at the same time they do not give any real information about the physical content of the phenomenon, about those things which really happen inside the atom. Of course, one can finally understand it, one can say: in quantum theory, in order to calculate an anomalous Zeeman effect we have to solve a perturbation problem, represented by a "secular" determinant. Such a "secular" determinant means a set of several linear equations with several unknowns. Now two coupled oscillators are just the same thing, they also mean several linear equations with several unknowns and therefore finally one can understand that these two theories are identical in their formalism although the physical content is extremely different.

The real success of these phenomenological attempts was on a slightly different line. In the course of time we tried in many cases to compare the formulae we got from the experiments with the Bohr theory. Something very funny happened. It was never possible to reproduce any of these formulae exactly by the Bohr theory. But still we got from the Bohr theory formulae which were somehow similar to the real formulae in the sense that, for instance, in the Bohr theory we would expect the square of the angular momentum, while empirically we had J (J + 1). Nowadays such a result is quite obvious because these are just representations of groups. But at that time this was a very strange result and it meant that somehow the Bohr theory was right and somehow it was quite incorrect, and we really did not know what to do about it. Because after all the quantum number, for instance, of the angular momentum was just defined as the value of the angular momentum, and it was practically impossible from classical physics that an expression like

$$\sqrt{J(J + 1)}$$

should emerge. We were quite upset about these results and at the same time we studied with extreme interest the recent papers of Bohr.

Bohr had just at that time published his papers on the periodic table of elements and we learned the very complicated structures of all the elements with ten or twenty or thirty electrons being in different orbits and we could not understand how Bohr could have obtained these results. We felt that he must have been an infinitely clever mathematician to solve such horrible problems of classical astronomy. We knew that even the problem of the three bodies had not been solved by the best astronomers, and there was now Neils Bohr, who could even solve problems with 30 electrons or something like that.

Bohr's conjecture

After two years of study, in the summer of 1922, Sommerfeld asked me whether I would be willing to follow him to a meeting at which Bohr would present his theory in Göttingen. These days in Göttingen we now always refer to as the "Bohr-festival". There for the first time I learned how a man like Bohr worked on problems of atomic physics. When Bohr had given two of his lectures I dared once in a discussion to utter some criticism; I just mentioned some doubts, whether the formulae of Kramers which he had written on the blackboard could be exact. I knew from our discussions in Munich that we always get formulae which are half exact, which are partly right and partly not right so I felt that it was never too certain. Bohr was very kind and in spite of the fact that I was a very young student, he asked me for a long walk on the Hainberg near Göttingen to discuss the problem. I feel it was then that I felt I really learned what it means to work on an entirely new field in theoretical physics. The first, for me quite shocking experience was that Bohr had calculated nothing. He had just guessed his results. He knew the experimental situation in chemistry, he knew the valencies of the various atoms and he knew that his idea of the quantization of the orbits or rather his idea of the stability of the atom to be explained by the phenomenon of quantization, fitted somehow with the experimental situation in chemistry. On this basis he simply guessed what he then gave us as his results. I asked him whether he really believed that one could derive these results by means of calculations based on classical mechanics. He said "Well, I think that those classical pictures which I draw of the atoms are just as good as classical pictures can be" and he explained it in the following way. He said "we are now in a new field of physics, in which we know that the old concepts probably don't work. We see that they don't work, because otherwise atoms wouldn't be stable. On the other hand when we want to speak about atoms, we must use words and these words can only be taken from the old concepts, from the old language. Therefore we are in a hopeless dilemma, we are like sailors coming to a very far away country. They don't know the country and they see people whose language they have never heard, so they don't know how to communicate. Therefore, so far as the classical concepts work, that is, so far as we can speak about the motion of electrons, about their velocity, about their energy, etc., I think that my pictures are correct or at least I hope that they are correct, but nobody knows how far such a language goes".

This was a very new way of thinking to me and it changed my whole attitude towards physics. In Sommerfeld's institute it always had seemed obvious that one should calculate something and only by rigorous calculation could one get good results.

Now coming back to the question of the phenomenological theories, I had the impression from my conversation with Bohr that one should go away from all these classical concepts, one should not speak about the orbit of an electron. In spite of the fact that you could see a track of the electron in the cloud chamber, you should not speak about the velocity or position and so on; but of course if you abandon these words then you don't know what to do. So this was a very strange dilemma and an extremely interesting situation and the question was "What can we do in such a situation?"

After this conversation with Bohr very soon, I think half a year later, I went to Copenhagen, I worked together with Kramers on dispersion theory and again we found this funny situation, that those formulae which could be derived from the Bohr theory were almost correct, but not really correct. One gradually acquired a kind of habit how to handle such formulae, how to translate from classical physics into these phenomenological formulae. One had the impression already that finally there must be some kind of quantum mechanics which has to replace classical mechanics. Quantum mechanics may be not too different from classical mechanics, but still it must use very different concepts.

Now in this situation it has often been said that it may be a step in the right direction to introduce into the theory only quantities that can be observed. Actually that was a very natural idea in this connection, because one saw that there were frequencies and amplitudes; and these frequencies and amplitudes could in classical theory somehow replace the orbit of the electron. A whole set of them means a Fourier series and a Fourier series describes an orbit. Therefore it was natural to think that one should use these sets of amplitudes and frequencies instead of the orbit.

When I came back from Copenhagen to Göttingen I decided that I should again try to do some kind of guesswork there, namely to guess the intensities in the hydrogen spectrum. The Bohr theory didn't work well for these intensities. But why should it not be possible to guess them? That was early in the summer of 1925 and I failed completely. The formulae got too complicated and there was no hope to get out anything. At the same time I also felt, if the mechanical system would be simpler, then it might be possible just to do the same thing as Kramers and I had done in Copenhagen and to guess the amplitudes. Therefore I turned from the hydrogen atom to the anharmonic oscillator, which was a very simple model. Just then I became ill and went to the island of Heligoland to recover. There I had plenty of time to do my calculations. It turned out that it really was quite simple to translate classical mechanics into quantum mechanics. But I should mention one important point. It was not sufficient simply to say "let us take some frequencies and amplitudes to replace orbit quantities" and use a kind of calculation which we had already used in Copenhagen and which later turned out to be equivalent to matrix multiplication.

It was quite clear that if one only did that, then one would have a scheme which was much more open than classical theory. Of course, classical theory would be included and quantum theory also would be included, but it was much too undefined and one had to add extra conditions.

It turned out that one could replace the quantum conditions of Bohr's theory by a formula which was essentially equivalent to the sum rule of Thomas and Kuhn. By adding such a condition one all of a sudden got into a consistent scheme. One could see that this set of assumptions worked, one could see that the energy was constant and so on. I was, however, not able to get a neat mathematical scheme out of it. Very soon afterwards both Born and Jordan in Göttingen and Dirac in Cambridge were able to invent a perfectly closed mathematical scheme; Dirac with very ingenious new methods on q numbers and Born and Jordan with more conventional methods on matrices.

Einstein on theory and observation

I don't want to speak about details now but rather about the interpretation of the details in the sense of asking "what kind of philosophy was the most important part in this development". To begin with I thought it was probably the idea of introducing only observable quantities. But when I had to give a talk about quantum mechanics in Berlin in 1926, Einstein listened to the talk and corrected this view.

Einstein asked me to come to his apartment and discuss the matters with him. The first thing he asked me was: "What was the philosophy underlying your kind of very strange theory? The theory looks quite nice, but what did you mean by only observable quantities?" I told him that I did not believe any more in electronic orbits, in spite of the tracks in a cloud chamber. I felt that one should go back to those quantities which really can be observed and I also felt that this was just the kind of philosophy which he had used in relativity; because he also had abandoned absolute time and introduced only the time of the special coordinate system and so on. Well, he laughed at me and then he said "but you must realize that it is completely wrong". I answered: "but why, is it not true that you have used this philosophy?" "Oh yes," he said, "I may have used it, but still it is nonsense!"

Einstein explained to me that it was really the other way around. He said "whether you can observe a thing or not depends on the theory which you use. It is the theory which decides what can be observed". His argument was like this: "Observation means that we construct some connection between a phenomenon and our realization of the phenomenon. There is something happening in the atom, the light is emitted, the light hits the photographic plate, we see the photographic plate and so on and so on. In this whole course of events between the atom and your eye and your consciousness you must assume that everything works as in the old physics. If you change the theory concerning this sequence of events then of course the observation would be altered". So he insisted that it was the theory which decides about what can be observed. This remark of Einstein was very important for me later on when Bohr and I tried to discuss the interpretation of quantum theory, and I shall come to that point later.

A few words more in connection with my discussion with Einstein. Einstein had pointed out to me that it is really dangerous to say that one should only speak about observable quantities. Every reasonable theory will, besides all things which one can immediately observe, also give the possibility of observing other things more indirectly. For instance, Mach himself had believed that the concept of the atom was only a point of covenience, a point of economy thinking, he didn't believe in the reality of the atoms. Nowadays everybody would say that this is nonsense, that it is quite clear that the atoms really exist. I also feel that one cannot gain anything by claiming that it is only a convenience of our thinking to have the atoms – though it may be logically possible. These were the points which Einstein raised. In quantum theory it meant, for instance, that when you have quantum mechanics then you cannot only observe frequencies and amplitudes, but for instance, also probability amplitudes, probability waves and so on, and these, of course, are quite different objects.

I should also add that when one has invented a new scheme which concerns certain observable quantities, then of course, the decisive question is: which of the old concepts can you really abandon? In the case of quantum theory it was more or less clear that you could abandon the idea of an electronic orbit.

Stability of laminar flow

But let me now leave this problem of phenomenological theories and come to the opposite question: what is the use of exact mathematical schemes? Perhaps you know that I am not at all fond of rigorous mathematical methods, and I would like to give a few reasons for this attitude. During these years before the formulation of quantum mechanics, I had to do my doctoral thesis. Since Sommerfeld was a good teacher, he felt that I should not always work on atomic theory; he told me "It is not good always to walk in the mud, you should really do decent mathematical work in theoretical physics".

So he suggested a hydrodynamical problem. I should calculate the stability of laminar flow. He had himself written a paper on this subject. The laminar flow between a resting and a moving wall had been treated by one of his pupils. Sommerfeld was not satisfied. This pupil, Hopf, had not been able to find a limit of stability. Experimentally, everybody knows that when the velocity gets too high then the laminar flow of a liquid goes over into a turbulent flow, eddies are created statistically, and this looks like a phenomenon of instability. Therefore it should be possible to calculate such a limit of stability. Sommerfeld suggested that I should calculate the stability of a stream of water between two fixed walls. This was my doctoral thesis and I got a nice result, as I believed, namely that there was a limit of stability. At a certain Reynolds number, in agreement with the experiment, the flow became unstable and one got turbulent motion.

Sequel after twenty years

Well, I got my degree on this paper; but one year later a very good mathematician, Noether, published another paper in which he proved by very rigorous mathematical methods that this problem which I had treated had no unstable solution; the flow should be stable everywhere. That was of course a very sad result, especially considering my degree, and I always hoped that I could disprove this paper of Noether. Unfortunately I was not able to disprove it; I only had to hope for the experiments, because experimentally there was certainly a limit of stability. This problem actually took many years before it became quite clear, and I may just mention a few steps. Five years later Tollmien treated a different kind of flow and actually got a limit of stability; he could argue that his was a different problem from Noether, so that the mathematical arguments of Noether did not apply. Then in 1944, that is 20 years after my doctoral thesis, in America Dryden and collaborators made very accurate measurements of the laminar flow between two walls and the transition to turbulence. They found that really the calculations which I had made agreed well with the experiment. Lin at M.I.T.* took up the problem and confirmed the old results by new and better methods. Still some mathematicians didn't believe it, we had long discussions on this problem in 1950 at M.I.T. and then von Neumann decided that one should use one of the electronic computers for the problem. So the

* Massachusetts Institute of Technology.

biggest computer at that time finally was used for killing the problem, and it turned out that the old approximate calculations of my thesis were off the correct values by not more than about 20%, and the question was now "what about this rigorous mathematical paper". Well, the trouble is, I think, that even now nobody knows what the mistake in the paper was.

Finding a mathematical mistake

But there was another case in which one knows where the mistake was. That was when Edward Teller came to my institute in Leipzig in 1928 or so, and he wanted to do a doctoral thesis. I didn't give him a problem on turbulence, because at that time already quantum mechanics was decent physics and so I suggested that he should be interested in the H-molecule, two protons and one electron. I told him that one of Bohr's pupils, Burrau, had just published a good paper on the normal state of this molecule ion, and had found a good value for the binding energy in agreement with the experiment. Teller should try for the excited states of the molecule.

A few weeks afterwards Teller came to my room, telling me that just recently a new paper by Wilson had appeared, using very good mathematical methods, much better than Burrau's, quite rigorous mathematics, and Wilson had been able to prove that the normal state of H did not exist. Well, again this was a rather sad result and I told Teller that it must be wrong, because after all, the molecule does exist and what can we do about it. But Teller said, "Wilson's mathematics is so good, you can't say anything against it". So Teller and I quarrelled quite a lot about it and after, I think, about two months or so Teller actually found the error in Wilson's paper; and it was quite an interesting mistake. The error was this: The mathematical methods were actually excellent, but Wilson had argued "we know that the Schrödinger function at far distance from the two centres must go to zero; this is correct. Therefore, our analytical function must be regular and have a zero at infinity" – which was wrong, because it was sufficient that it should go to zero on the real axis, not on the imaginary axis. Well, this is just the kind of mistake which one can make and I hope that Noether has made a similar mistake in the turbulence problem, but that I don't know.

Rigorous and dirty mathematics

I think that you understand now why I am always a bit sceptical of rigorous mathematical methods.

Perhaps I should give a more serious reason for that: When you try too much for rigorous mathematical methods, you fix your attention on those points which are not important from the physics point and thereby you get away from the experimental situation. If you try to solve a problem by rather dirty mathematics, as I have mostly done, then you are forced always to think of the experimental situation; and whatever formulae you write down, you try to compare the formulae with reality and thereby, somehow, you get closer to reality than by looking for the rigorous methods. But this may, of course, be different for different people.

Let us now come back to quantum mechanics, and to that part of the development of a new theory that has always seemed to me the most fascinating part. When you get into such a new field, the trouble is, that with phenomenological methods you are bound always to use the old concepts; because you have no other concepts, and making theoretical connections means then applying old methods to this new situation. Therefore the decisive step is always a rather discontinuous step. You can never hope to go by small steps nearer and nearer to the real theory; at one point you are bound to jump, you must really leave the old concepts and try something new and then see whether you can swim or stand or whatever else, but in any case you can't keep the old concepts. This happened in quantum mechanics in the following way: first we had the mathematical scheme, and then, of course, we had to try to use a reasonable language in connection with it. Finally we could ask: what concepts does this mathematical scheme imply and how do we have to describe nature?

Abandoning old concepts

The most difficult part in this stage of development is abandoning some of the important old concepts. Any good physicist would be willing to acquire new concepts but even the best physicists are sometimes quite unwilling to leave some of the old and apparently safe concepts. This feeling that one cannot go away from the old concepts was also very strong in the development of quantum mechanics. You know that it has been very strong in the development of relativity, and even nowadays there are papers appearing here and there in which people just refuse to understand the theory of special relativity. They cannot understand it because they are not able to go away from the old concept of "simultaneous events". In quantum theory the same thing happened to some extent in the discussions about wave mechanics of Schrödinger and quantum mechanics. I remember one lecture of Schrödinger and the discussion afterwards in the summer of 1926. Perhaps I should mention it, certainly not to criticize Schrödinger, who was a first class physicist, but just to show how extremely difficult it is to get away from old concepts. Schrödinger had given a lecture on wave mechanics, he had been invited by Sommerfeld and there was also Wilhelm Wien, who was an experimental physicist; at that time the theory of Bohr was not at all generally recognised as a good theory. The experimental physicists, for instance, in Munich disliked all this game of quantum terms and quantum jumps, they called it atomystic, that is mysticism of the atom, and they felt that it was so much unlike classical physics that it was really not to be taken seriously. Therefore Wilhelm Wien was extremely glad to hear from Schrödinger his new interpretation.

You know that Schrödinger for a time believed that he could use wave mechanics with the same kind of concepts as Maxwell theory. He assumed that the matter waves are just three-dimensional waves in space and time like electromagnetic waves, and therefore the eigenvalue of an energy was really the eigenvalue of a vibration and not an energy. Thereby he believed that he could avoid all kinds of quantum jumps and all the rest of what he called mysticism. After Schrödinger's lecture I took part in the discussion and argued that I felt that by such an interpretation one could not even understand Planck's law. Because after all, Planck's law was based on real quantum theory, on the discontinuous changes of energy and so on. Then Wien was so angry about this remark that he said "Well young man, we understand that your are sorry that now quantum mechanics and quantum jumps and all the rest should be forgotten, but you will see that Schrödinger will solve all these problems very soon".

I just mentioned this episode to show how strong the feelings can be among physicists about such matters. Of course, I was completely unsuccessful in convincing either Wien or Schrödinger; but the result was that Bohr invited Schrödinger to Copenhagen. Schrödinger, in September of 1926, came to Copenhagen. Bohr, a very kind and fine man and most amiable in every way, could sometimes be almost fanatical. I remember, wherever Schrödinger stood, Bohr would also stand there, saying "But Schrödinger you must understand, you must really". After two days Schrödinger became ill. He had to go to bed and Mrs. Bohr would bring cake and tea and so on, but Bohr would sit at the bedside: "But Schrödinger, you must understand". After this time Schrödinger at least understood that it was more difficult with the interpretation of quantum theory than he had thought.

Also in Copenhagen we were not yet too happy about the interpretation because we felt that in the atom it seemed all right to abandon the concept of an electronic orbit. But what about in a cloud chamber? In a cloud chamber you see the electron moving along the track; is this an electronic orbit or not?

Quantum theory understood

Bohr and I discussed these problems many, many nights and we were frequently in a state of despair. Bohr tried more in the direction of the duality between waves and particles; I preferred to start from the mathematical formalism and to look for a consistent interpretation. Finally Bohr went to Norway to think alone about the problem and I remained in Copenhagen. Then I remembered Einstein's remark in our discussion. I remembered that Einstein had said that "It is the theory which decides what can be observed". From there it was easy to turn around our question and not to ask "How can I represent in quantum mechanics this orbit of an electron in a cloud chamber?" but rather to ask "Is it not true that always only such situations occur in Nature, even in a cloud chamber, which can be described by the mathematical formalism of quantum mechanics?" By turning around I had to investigate what can be described in this formalism; and then it was very easily seen, especially when one used the new mathematical discoveries of Dirac and Jordan about transformation theory, that one could not describe at the same time the exact position and the exact velocity of an electron; one had these uncertainty relations. In this way things became clear. When Bohr returned to Copenhagen, he had found an equivalent interpretation with his concept of complementarity, so finally we all agreed that now we had understood quantum theory.

Einstein's fictitious experiments

Again, we met a difficult situation in 1927 when Einstein and Bohr discussed these matters at the Solvay Conference. Almost every day the sequence of events was the following: We all lived in the same hotel. In the morning for breakfast Einstein would appear and tell Bohr a new fictitious experiment in which he could disprove the uncertainty relations and thereby our interpretation of quantum theory. Then Bohr, Pauli and I would be very worried, we would follow Bohr and Einstein to the meeting and would discuss this problem all day. But at night for dinner usually Bohr had solved the problem and he gave the answer to Einstein, so then we felt that everything was alright and Einstein was a bit sorry about it and said he would think about it. Next morning he would bring a new fictitious experiment, again we had to discuss it, etc. This went on for quite a number of days and at the end of the conference the Copenhagen physicists had the feeling that they had won the battle and that actually Einstein could not make any real objection. I think the most splendid argument of Bohr was that he once used the Theory of General Relativity to disprove Einstein. Einstein had invented an experiment in which the weight of some machinery was to be determined by gravitation and so Bohr had to invoke the Theory of General Relativity to show that the uncertainty relations were correct. Bohr succeeded and Einstein could not raise any objection.

Electrons and the nucleus

Now I come to more recent developments. Perhaps I should, before I come to relativistic quantum theory, say a few words about nuclear physics. The only point I want to make here again is that it is much easier to accept new concepts than to abandon old ones. Actually, when the neutron was discovered by Chadwick in 1932, I think, it was almost trivial to say that the nucleus consists of protons and neutrons, but it was not quite so trivial to say that there are no electrons in the nucleus. The decisive point of those papers that I wrote about the structure of the nucleus was not that the nucleus consisted of protons and neutrons, but that in apparent contradiction to experiment there were no electrons in the nucleus. Everybody up to that time had assumed that there must be electrons in the nucleus, because sometimes they come out, and it was rather odd to say that they have not been in the nucleus before. Of course, the idea was that the short-range forces between neutron and proton somehow might have to create electrons in the nucleus. Anyway it seemed to be a good approximation to assume that such light particles cannot exist in the nucleus. I remember that I have been criticized very strongly for this assumption by extremely good physicists. I got one letter saying that it was really a scandal to assume that there were no electrons in the nucleus because one could just see them coming out; I would bring complete disorder into physics by such unreasonable assumptions and they could not understand my attitude. I just mentioned this small event, because it is really difficult to go away from something which seems so natural and so obvious that everybody had always accepted it. I think the greatest effort in the development of theoretical physics is always necessary at those points where one has to abandon old concepts.

Changing the outlook of atomic physics

May I now turn to the problem of the elementary particles. I think that really the most decisive discovery in connection with the properties or the nature of elementary particles was the discovery of antimatter by Dirac. That was an entirely new feature which apparently had to do with relativity, with the replacement of the Galilei group by the Lorentz group. I believe that this discovery of particles and antiparticles by Dirac has changed our whole outlook on atomic physics completely. I do not know whether this change was realized at once at that time, probably it has been accepted only gradually; but I would like to explain why I consider it so fundamental.

We know from quantum theory that, for instance, a hydrogen molecule may consist of two hydrogen atoms or of one positive hydrogen ion and one negative hydrogen ion. Generally one can say that every state consists virtually of all possible configurations by which you can realize the same kind of symmetry. Now as soon as one knows that one can create pairs according to Dirac's theory, then one has to consider an elementary particle as a compound system; because virtually it could be this particle plus a pair or this particle plus two pairs and so on, and so all of a sudden the whole idea of an elementary particle has changed. Up to that time I think every physicist had thought of the elementary particles along the line of the philosophy of Democritus, namely by considering these elementary particles are unchangeable units which are just given in nature and are just always the same thing, they never change, they never can be transmuted into anything else. They are not dynamical systems, they just exist in themselves.

After Dirac's discovery everything looked different, because now one could ask, why should a proton be only a proton, why should a proton not sometimes be a proton plus a pair of one electron and one positron and so on. This new aspect of the elementary particle being a compound system at once looked to me as a great challenge. When later I worked together with Pauli on quantum electrodynamics, I always kept this problem in my mind.

Pair creation

The next step in this direction was the idea of multiple production of particles. If two particles collide, then pairs can be created; then there is no reason why there should only be one pair; why should there not be two pairs. If only the energy is high enough one could eventually have any number of particles created by such an event, if the coupling is strong enough. Thereby the whole problem of dividing matter had come into a different light. So far one had believed that there are just two alternatives. Either you can divide matter again and again into smaller and smaller bits or you cannot divide matter up to infinity and then you come to smallest particles. Now all of a sudden we saw a third possibility: we can divide matter again and again but we never get to smaller particles because we must create particles by energy, by kinetic energy, and since we have pair creation this can go on forever. So it was a natural but paradoxical concept to think of the elementary particle as a compound system of elementary particles. Of course then the problem arose: "What kind of mathematical scheme can describe such a situation?"

At that time one knew from Dirac's theory of radiation and from the attempts of Pauli, Weisskopf and myself that one had great difficulties in avoiding infinities in quantum electrodynamics and, more generally, in quantum field theory with interaction. I agree completely with Dirac in disliking infinities in the sense that if you introduce infinity in physics, you just talk nonsense, it cannot be done. Therefore I tried to think of mathematical schemes in which you can avoid infinities from the very beginning. Again, I remembered the old story of the observable quantities and therefore I felt that it was probably useful to ask "what can we really observe in a collision between elementary particles", and so it was natural to come to the S-matrix and to say that the S-matrix or scattering matrix is a rational basis for a theory.

Again of course it is considerably easier to go this first step, namely to say that such and such things can be observed, than to go the next step and to narrow down the assumptions. But finally you have to make new assumptions and end up with saying "Such and such things cannot be observed anymore". So the question was now "How can we narrow down the concept of the S-matrix in order to get something which is really workable, in which we can define what we mean, in which we can formulate natural laws". Well, at that time I had learned again from Dirac that one could perhaps work with a field theory making use of an indefinite metric in Hilbert space. I knew of course that Pauli had criticized that very strongly, as he always did – and very often successfully – namely by saying that if one had an indefinite metric in Hilbert space it meant negative probabilities, and therefore such a scheme would not work.

It was natural to think of the possibility that for the asymptotic region of course you must have positive probabilities, therefore asymptotically you must have a unitary S-matrix; but at the same time it would be allowed to go away locally from this concept of probability and say "Locally we may have negative probabilities because locally we cannot measure anything in the same way as in the asymptotic region". The concept of probability may fail when we get below a certain "universal length". Therefore I tried to narrow the scheme down by saying "There shall be local field operators but these operators may work in a Hilbert space which does not have an ordinary metric but may have an indefinite metric." The advantage of this scheme was that one could actually avoid infinities but, of course, at a very high price, namely at the price of loosing the definite metric in Hilbert space. On the other hand, by that time the whole scheme already looked rather convincing to me, because the experiments in the meantime had proved that there was actually multiple production of particles.

Phenomenon established

That had in fact been a rather controversial subject for more than ten years, because there were, of course, these cosmic-ray showers which everybody had known since 1936 or so. But these showers could be very well explained by the cascade theory of Bhabha and Heiter; so there was no proof for multiple production of particles. It was not before about 1950 that one really could get very good evidence for the existence of this multiple production. But since this phenomenon was now well established, I felt that one could go on in the same direction and therefore I tried to formulate a kind of field theory. I thought that for the mathematical scheme the model of Lee could give some help, but of course I was very well aware of the fact that in field theory we had no rigorous mathematical scheme. I felt that it might be sufficient for the time being to look for a mathematical scheme which somehow fitted the experimental situation.

To begin with, we did not know of any good field equation which could represent the actual situation and the experiments. But then in 1956, after a lecture I had given at CERN,* I met Pauli and we discussed the new possibilities. We had learned from Lee about the nonconservation of parity in β-decay, and following up this line of ideas we came to a field equation which turned out to contain the SU2 group, namely the Isospin. Pauli was more enthusiastic about this possibility than I have ever seen him. I got letters from him saying that now a new dawn of physics has begun and all our difficulties will disappear very soon and so on. I always had to stop him and say "Well, that is not so easy." But he was so excited about it and full of energy and enthusiasm, his central interest was to work on these problems.

During this period I met him several times in Zurich, but then he had to go to the States. When he had to give lectures on these problems there he tried to rationalize his feelings and then he felt that he was not able to do it; he saw that the whole problem

* The European Centre for research in nuclear and particle physics in Geneva, Switzerland.

was much more complicated than he had hoped for. I should perhaps mention in this connection that the most essential idea that Pauli had contributed to our common paper was (in a somewhat preliminary form) the idea of a degeneracy of the ground state which later, in connection with the theorem of Goldstone, has played a considerable role in elementary particle physics.

Pauli's critical acumen

Pauli's whole character was different from mine. He was much more critical and he tried to do two things at once while I would think that this is really too difficult even for the best physicist. He tried first of all to be inspired by the experiments and to see in a kind of intuitive way how things are connected, and at the same time he tried to rationalize his intuitions and to find a rigorous mathematical scheme so that he really could prove everything that he said. Now this is, I think, simply too much, and therefore Pauli has through his whole life published much less than he could have published if he had abandoned one of these two postulates. Bohr had dared publish papers which he could not prove and which were right after all. Others have done a lot by rational methods and by good mathematics, but the two things together, that I think is too much for one man. Pauli was completely disappointed when he saw the difficulties and so he gave up in a rather sad way. He told me that he felt that his thinking was not strong enough any more and that he was not well at all; but he encouraged me even after he had withdrawn his approval for the publication. I should go on, he said, but he could not continue and as you know, unfortunately died half a year later. This was rather a sad end to my long friendship with Pauli and I can only say that I regret even now almost every day that I cannot have his very strong criticism which has helped so many times in my life in physics.

But let us come back now to the further developments in physics. I think we know more about the degeneracy of the ground state and perhaps most of you will know much more of the details and of the mathematical scheme than I do. I can only hope that the picture will remain a closed picture. I cannot doubt that it is possible to describe the whole spectrum of the elementary particles in the same way as we describe the spectrum of say the iron atom in quantum mechanics, that is by one unified natural law. This law of course will be a kind of summary, a resume of the many details which are being studied now.

MY GENERAL PHILOSOPHY

It may be tempting to add a prescription about how we should work in theoretical physics. This would, however, be very dangerous, because the prescription ought to be different for different physicists. Therefore I can only speak about the prescription that I have always used for myself. This was that one should not stick too much to one special group of experiments; one should rather try to keep in touch with all the developments in all the relevant experiments so that one should always have the whole picture in mind before one tries to fix a theory in mathematical or other languages.

I might perhaps describe this general philosophy by telling two quite different stories. When I was a boy, my grandfather, who was a handicraftsman and knew how to do practical things, once met me when I put a cover on a wooden box with books or so. He saw that I took the cover and I took a nail and I tried to hammer this one nail down to the bottom. "Oh", he said, "that is quite wrong what you do there, nobody can do it that way and it is a scandal to look at". I did not know what the scandal was, but then he said, "I will show you how you could do it". He took the cover and he took one nail, put it just a little bit through the cover into the box, and then the next nail a little bit, the third nail a little bit and so on until all the nails were there. Only when everything was clear, when one could see that all the nails would fit, then he would start to put the nails really into the box. So, I think this is a good description of how one should proceed in theoretical physics.

The other story concerns the discussions which Dirac and I had. Dirac often liked to say – and I always felt that it was a slight criticism – he felt, that one can only solve one difficulty at a time. This may be right, but it was not the way I looked at the problems. Then I remembered that Niels Bohr used to say "If you have a correct statement, then the opposite of a correct statement is of course an incorrect statement, a strong statement. But when you have a deep truth, then the opposite of a deep truth may again be a deep truth." Therefore I feel that it is perhaps not only a deep truth to say "You can only solve one difficulty at a time", but it may also be a deep truth to say "You can never solve only one difficulty at a time, you have to solve always quite a lot of difficulties at the same time" and with this remark perhaps I should close my talk.

METHODS IN THEORETICAL PHYSICS*

Paul Adrian Maurice Dirac

I shall attempt to give you some idea of how a theoretical physicist works – how he sets about trying to get a better understanding of the laws of nature.

One can look back over the work that has been done in the past. In doing so one has the underlying hope at the back of one's mind that one may get some hints or learn some lessons that will be of value in dealing with present-day problems. The problems that we had to deal with in the past had fundamentally much in common with the present-day ones, and reviewing the successful methods of the past may give us some help for the present.

One can distinguish between two main procedures for a theoretical physicist. One of them is to

*The text of this lecture is reprinted here with kind permission of the International Centre for Theoretical Physics, Trieste.

work from the experimental basis. For this, one must keep in close touch with the experimental physicists. One reads about all the results they obtain and tries to fit them into a comprehensive and satisfying scheme.

The other procedure is to work from the mathematical basis. One examines and criticizes the existing theory. One tries to pin-point the faults in it and then tries to remove them. The difficulty here is to remove the faults without destroying the very great successes of the existing theory.

There are these two general procedures, but of course the distinction between them is not hard-and-fast. There are all grades of procedure between the extremes.

Which procedure one follows depends largely on the subject of study. For a subject about which very little is known, where one is breaking quite new ground, one is pretty well forced to follow the procedure based on experiment. In the beginning, for a new subject, one merely collects experimental evidence and classifies it.

For example, let us recall how our knowledge of the periodic system for the atoms was built up in the last century. To begin with, one simply collected the experimental facts and arranged them. As the system was built up one gradually acquired confidence in it, until eventually, when the system was nearly complete, one had sufficient confidence to be able to predict that, where there was a gap, a new atom would subsequently be discovered to fill the gap. These predictions all came true.

In recent times there has been a very similar situation for the new particles of high-energy physics. They have been fitted into a system in which one has so much confidence that, where one finds a gap, one can predict that a particle will be discovered to fill it.

In any region of physics where very little is known, one must keep to the experimental basis if one is not to indulge in wild speculation that is almost certain to be wrong. I do not wish to condemn speculation altogether. It can be entertaining and may be indirectly useful even if it does turn out to be wrong. One should always keep an open mind receptive to new ideas, so one should not completely oppose speculation, but one must take care not to get too involved in it.

Cosmological speculation

One field of work in which there has been too much speculation is cosmology. There are very few hard facts to go on, but theoretical workers have been busy constructing various models for the universe, based on any assumptions that they fancy. These models are probably all wrong. It is usually assumed that the laws of nature have always been the same as they are now. There is no justification for this. The laws may be changing, and in particular quantities which are considered to be constants of nature may be varying with cosmological time. Such variations would completely upset the model makers.

With increasing knowledge of a subject, when one has a great deal of support to work from, one can go over more and more towards the mathematical procedure. One then has as one's underlying motivation the striving for mathematical beauty. Theoretical physicists accept the need for mathematical beauty as an act of faith. There is no compelling reason for it, but it has proved a very profitable objective in the past. For example, the main reason why the theory of relativity is so universally accepted is its mathematical beauty.

With the mathematical procedure there are two main methods that one may follow, (i) to remove inconsistencies and (ii) to unite theories that were previously disjoint.

SUCCESS THROUGH METHOD

There are many examples where the following of method (i) has led to brilliant success. Maxwell's investigation of an inconsistency in the electromagnetic equations of his time led to his introducing the displacement current, which led to the theory of electromagnetic waves. Planck's study of difficulties in the theory of black-body radiation led to his introduction of the quantum. Einstein noticed a difficulty in the theory of an atom in equilibrium in black-body radiation and was led to introduce stimulated emission, which has led to the modern lasers. But the supreme example is Einstein's discovery of his law of gravitation, which came from the need to reconcile Newtonian gravitation with special relativity.

In practice, method (ii) has not proved very fruitful. One would think that the gravitational and electromagnetic fields, the two long-range fields known in physics, should be closely connected, but Einstein spent many years trying to unify them, without success. It seems that a direct attempt to unify disjoint theories, where there is no definite inconsistency to work from, is usually too difficult, and if success does ultimately come, it will come in an indirect way.

Whether one follows the experimental or the mathematical procedure depends largely on the subject of study, but not entirely so. It also depends on the man. This is illustrated by the discovery of quantum mechanics.

Two men are involved, Heisenberg and Schrödinger. Heisenberg was working from the experimental basis, using the results of spectroscopy, which by 1925 had accumulated an enormous amount of data. Much of this was not useful, but some was, for example, the relative intensities of the lines of a multiplet. It was Heisenberg's genius that he was able to pick out the important things from the great wealth of information and arrange them in a natural scheme. He was thus led to matrices.

Schrödinger's approach was quite different. He worked from the mathematical basis. He was not well informed about the latest spectroscopic results, like Heisenberg was, but had the idea at the back of his mind that spectral frequencies should be fixed by eigenvalue equations, something like those that fix the frequencies of systems of vibrating springs. He had this idea for a long time, and was eventually able to find the right equation, in an indirect way.

Impact of relativity

In order to understand the atmosphere in which theoretical physicists were then working, one must appreciate the enormous influence of relativity. Relativity had burst into the world of scientific thought with a tremendous impact, at the end of a long and difficult war. Everyone wanted to get away from the strain of war and eagerly seized on the new mode of thought and new philosophy. The excitement was quite unprecedented in the history of science.

Against this background of excitement, physicists were trying to understand the mystery of the stability of atoms. Schrödinger, like everyone else, was caught up by the new ideas, and so he tried to set up a quantum mechanics within the framework of relativity. Everything had to be expressed in terms of vectors and tensors in space-time. This was unfortunate, as the time was not ripe for a relativistic quantum mechanics, and Schrödinger's discovery was delayed in consequence.

Schrödinger was working from a beautiful idea of de Broglie connecting waves and particles in a relativistic way. De Broglie's idea applied only to free particles, and Schrödinger tried to generalize it to an electron bound in an atom. Eventually he succeeded, keeping within the relativistic framework. But when he applied his theory to the hydrogen atom, he found it did not agree with experiment. The discrepancy was due to his not having taken the spin of the electron into account. It was not then known. Schrödinger subsequently noticed that his theory was correct in the nonrelativistic approximation, and he had to reconcile himself to publishing this degraded version of his work, which he did after some month's delay.

The moral of this story is that one should not try to accomplish too much in one go. One should separate the difficulties in physics one from another as far as possible, and then dispose of them one by one.

Heisenberg and Schrödinger gave us two forms of quantum mechanics, which were soon found to be equivalent. They provided two pictures, with a certain mathematical transformation connecting them.

I joined in the early work on quantum mechanics, following the procedure based on mathematics, with a very abstract point of view. I took the noncommutative algebra which was suggested by Heisenberg's matrices as the main feature for a new dynamics, and examined how classical dynamics could be adapted to fit in with it. Other people were working on the subject from various points of view, and we all obtained equivalent results, at about the same time.

Fruitful relaxation

I would like to mention that I found the best ideas usually came, not when one was actively striving for them, but when one was in a more relaxed state. Professor Bethe has told us how he got ideas on railway trains and often worked them out before the end of the journey. It was not like that with me. I used to take long solitary walks on Sundays, during which I tended to review the current situation in a leisurely way. Such occasions often proved fruitful, even though (or perhaps because) the primary purpose of the walk was relaxation and not research.

It was on one of these occasions that the possibility occurred to me of a connection between commutators and Poisson brackets. I did not then know very well what a Poisson bracket was, so was very uncertain of the connection. On getting home I found I did not have any book explaining Poisson brackets, so I had to wait impatiently for the libraries to open the following morning before I could verify the idea.

PAUL DIRAC

With the development of quantum mechanics one had a new situation in theoretical physics. The basic equations, Heisenberg's equations of motion, the commutation relations and Schrödinger's wave equation were discovered without their physical interpretation being known. With noncommutation of the dynamical variables, the direct interpretation that one was used to in classical mechanics was not possible, and it became a problem to find the precise meaning and mode of application of the new equations.

This problem was not solved by a direct attack. People first studied examples, such as the nonrelativistic hydrogen atom and Compton scattering, and found special methods that worked for these examples. One gradually generalized, and after a few years the complete understanding of the theory was evolved as we know it today, with Heisenberg's principle of uncertainty and the general statistical interpretation of the wave function.

The early rapid progress of quantum mechanics was made in a nonrelativistic setting, but of course people were not happy with this situation. A relativistic theory for a single electron was set up, namely Schrödinger's original equation, which was rediscovered by Klein and Gordon and is known by their name, but its interpretation was not consistent with the general statistical interpretation of quantum mechanics.

From tensors to spinors

As relativity was then understood, all relativistic theories had to be expressible in tensor form. On this basis one could not do better than the Klein–Gordon theory. Most physicists were content with the Klein–Gordon theory as the best possible relativistic quantum theory for an electron, but I was always dissatisfied with the discrepancy between it and general principles, and continually worried over it till I found the solution.

Tensors are inadequate and one has to get away from them, introducing two-valued quantities, now called spinors. Those people who were too familiar with tensors were not fitted to get away from them and think up something more general, and I was able to do so only because I was more attached to the general principles of quantum mechanics than to tensors. Eddington was very surprised when he saw the possibility of departing from tensors. One should always guard against getting too attached to one particular line of thought.

The introduction of spinors provided a relativistic theory in agreement with the general principles of quantum mechanics, and also accounted for the spin of the electron, although this was not the original intention of the work. But then a new problem appeared, that of negative energies. The theory gives symmetry between positive and negative energies, while only positive energies occur in nature.

As frequently happens with the mathematical procedure in research, the solving of one difficulty leads to another. You may think that no real progress is then made, but this is not so, because the second difficulty is more remote than the first. It may be that the second difficulty was really there all the time, and was only brought into prominence by the removal of the first.

This was the case with the negative energy difficulty. All relativistic theories give symmetry between positive and negative energies, but previously this difficulty had been overshadowed by more crude imperfections in the theory.

The difficulty is removed by the assumption that in the vacuum all the negative energy states are filled. One is then led to a theory of positrons together with electrons. Our knowledge is thereby advanced one stage, but again a new difficulty appears, this time connected with the interaction between an electron and the electromagnetic field.

When one writes down the equations that one believes should describe this interaction accurately and tries to solve them, one gets divergent integrals for quantities that ought to be finite. Again this difficulty was really present all the time, lying dormant in the theory, and only now becoming the dominant one.

On the wrong track?

If one deals classically with point electrons interacting with the electromagnetic field, one finds difficulties connected with the singularities in the field. People have been aware of these difficulties from the time of Lorentz, who first worked out the equations of motion for an electron. In the early days of the quantum mechanics of Heisenberg and Schrödinger, people thought these difficulties would be swept away by the new mechanics. It now became clear that these hopes would not be fulfilled. The difficulties reappear in the divergencies of quantum electrodynamics, the quantum theory of the interaction of electrons and the electromagnetic field. They are modified somewhat by the infinities associated with the sea of negative-energy electrons, but they stand out as the dominant problem.

The difficulty of the divergencies proved to be a

very bad one. No progress was made for twenty years. Then a development came, initiated by Lamb's discovery and explanation of the Lamb shift, which fundamentally changed the character of theoretical physics. It involved setting up rules for discarding the infinities, rules which are precise, so as to leave well-defined residues that can be compared with experiment. But still one is using working rules and not regular mathematics.

Most theoretical physicists nowadays appear to be satisfied with this situation, but I am not. I believe that theoretical physics has gone on the wrong track with such developments and one should not be complacent about it. There is some similarity between this situation and the one in 1927, when most physicists were satisfied with the Klein–Gordon equation and did not let themselves be bothered by the negative probabilities that it entailed.

We must realize that there is something radically wrong when we have to discard infinities from our equations, and we must hang on to the basic ideas of logic at all costs. Worrying over this point may lead to an important advance. Quantum electrodynamics is the domain of physics that we know most about, and presumably it will have to be put in order before we can hope to make any fundamental progress with the other field theories, although these will continue to develop on the experimental basis.

Let us see what can be done with putting the present quantum electrodynamics on a logical footing. We must keep to the standard practice of neglecting only quantities which one can believe to be small, even though the grounds for this belief may be rather shaky.

In order to handle infinities, we must refer to a process of cut-off. We must do this in mathematics whenever we have a series or an integral which is not absolutely convergent. When we have introduced a cut-off, we may proceed to make it more and more remote and go to a limit, which then depends on the method of cut-off. Alternatively, we may keep the cut-off finite. In the latter case, we must find quantities that are insensitive to the cut-off.

The divergencies in quantum electrodynamics come from the high-energy terms in the energy of interaction between the particles and the field. The cut-off thus involves introducing an energy, g say, beyond which the interaction energy terms are omitted. It is found that we cannot make g tend to infinity without destroying the possibility of solving the equations logically. We have to keep a finite cut-off.

The relativistic invariance of the theory is then destroyed. This is a pity, but it is a lesser evil than a departure from logic would be. It results in a theory which cannot be valid for high-energy processes, processes involving energies comparable with g, but we may still hope that it will be a good approximation for low-energy processes.

On physical grounds we should expect to have to take g to be of the order of a few hundred Mev, as this is the region where quantum electrodynamics ceases to be a self-contained subject and the other particles of physics begin to play a role. This value for g is satisfactory for the theory.

Working with a finite cut-off, we have to search for quantities which are not sensitive to the precise mode and value of the cut-off. We then find that the Schrödinger picture is not a suitable one. Solutions of the Schrödinger equation, even the one describing the vacuum state, are very sensitive to the cut-off. But there are some calculations that one can carry out in the Heisenberg picture that lead to results insensitive to the cut-off.

One can deduce in this way the Lamb-shift and the anomalous magnetic moment of the electron.

The results are the same as those obtained some twenty years ago by the method of working rules with discard of infinities. But now the result can be obtained by a logical process, following standard mathematics in which only small quantities are neglected.

As we cannot now use the Schrödinger picture, we cannot use the regular physical interpretation of quantum mechanics involving the square of the modulus of the wave function. We have to feel our way towards a new physical interpretation which can be used with the Heisenberg picture. The situation for quantum electrodynamics is rather like that for elementary quantum mechanics in the early days when we had the equations of motions but no general physical interpretation.

A feature of the calculations leading to the Lamb shift and anomalous magnetic moment should be noted. One finds that the parameters m and e denoting the mass and charge of the electron in the starting equations are not the same as the observed values for these quantities. If we keep the symbols m and e to denote the observed values, we have to replace the m and e in the starting equations by $m + \delta m$ and $e + \delta e$, where δm and δe are small corrections which can be calculated. This procedure is known as renormalization.

Such a change in the starting equations is permitted. We can take any starting equations we like, and then develop the theory by making deductions from them. You might think that the work of the theoretical physicist is easy if he can make any starting assumptions he likes, but the difficulty arises because he needs the same starting assumptions for all the applications of the theory. This very strongly restricts his freedom. Renormalization is permitted because it is a simple change which can be applied universally whenever one has charged particles interacting with the electromagnetic field.

There is a serious difficulty still remaining in quantum electrodynamics, connected with the self-energy of the photon. It will have to be dealt with by some further change in the starting equations, of a more complicated kind than renormalization.

The ultimate goal is to obtain suitable starting equations from which the whole of atomic physics can be deduced. We are still far from it. One way of proceeding towards it is first to perfect the theory of low-energy physics, which is quantum electrodynamics, and then try to extend it to higher and higher energies. However, the present quantum electrodynamics does not conform to the high standard of mathematical beauty that one would expect for a fundamental physical theory, and leads one to suspect that a drastic alteration of basic ideas is still needed.